CONSTRUCTIVE APPROXIMATION

Subscription Information

Constructive Approximation is published quarterly by Springer-Verlag New York Inc.

NORTH AMERICA: Institutional subscription rate: $114.00, including postage and handling. Subscriptions are entered with prepayment only. Please mail your order to: Springer-Verlag New York Inc., Service Center Secaucus, 44 Hartz Way, Secaucus, NJ 07094, USA. Tel. (201) 348-4033, Telex 12 59 94.

ALL OTHER COUNTRIES: Subscription rate: DM 264 plus postage and handling. Please place your order with your bookdealer or mail directly to: Springer-Verlag, Heidelberger Platz 3, D-1000 Berlin 33, Germany. Tel. (030) 82071, Telex 18 33 19.

BACK VOLUMES AND MICROFORM EDITIONS: Back volumes are available on request from the publisher. Microform editions are available from: University Microfilms International, 300 N. Zeeb Road, Ann Arbor, MI 48106.

CHANGE OF ADDRESS: Please inform the publisher of your new address 6 weeks before you move. All communications should include both old and new addresses (with Zip codes) and should be accompanied by a mailing label from a recent issue.

ISSN: 0176-4276

Copyright

ISBN 978-1-4899-6816-6 ISBN 978-1-4899-6886-9 (eBook)
DOI 10.1007/978-1-4899-6886-9

Constr. Approx. (1989) 5: 1–2

CONSTRUCTIVE APPROXIMATION

Preface

This special issue marks an exciting moment in the history of mathematics.

Let me explain. I believe that mathematics holds the key to the control of the massive amounts of data which mark the new socioeconomic era. Technologies come and others go, directly in accord with their abilities to handle more information faster. Companies rise and fall with these technologies.

Concurrent with the publication of this special issue there has appeared a new technology for processing and transmitting images. A single digital image of size 1024×1024 with 24 bits/pixel requires three megabytes of computer storage, and takes half an hour to communicate by telephone. This communication time can be reduced to seconds by replacing the images by fractals. The approximation of data by fractals is the central thrust of this special issue.

The theorems herein are blueprints for new technological devices. Almost any one can be interpreted in a discrete setting, converted to digital algorithms, translated to software, optimized, then converted to hardware designs, and passed to a silicon foundry. The product may be part of a new communications device. The difference between now and earlier eras is the speed with which the technology transfer can take place: from mathematics to a component inside a commercially available television set can take place in less than three years. Mathematicians hold the key to this process.

Fractal geometry is a new focus for mathematical research. This research draws both on classical analysis and computer graphics and upon interplay with the rapidly developing applications. As to what the body of knowledge which we might now want to call "fractal geometry" will finally look like is not clear: so much is going on, so many diverse branches of study involve fractals in one way or another, and yet contain great creativity beyond their fractal theme, that it is impossible yet to say how the area will be defined eventually. One thing is clear however: an important part of the area will belong to approximation theory. It is the aim of this special issue to try and support this assertion and to go some way toward introducing, if only by illustration, the main ingredients and directions which will belong to Fractal Approximation Theory.

In the following paragraphs I give some of the questions that an "approximation theory" must address, and I say how this issue addresses them.

What objects will Fractal Approximation Theory be concerned with approximating? Complicated subsets of $\mathbf{R}^2$, and finite Borel measures supported on these subsets. An example of the former is provided by wriggly, nondifferentiable functions, such as might occur in modeling a time series of temperatures in a jet-engine exhaust. Another example is a subset of $\mathbf{R}^2$ which represents a picture of a fern, black on a white background. Measures on $\mathbf{R}^2$ occur in great diversity in the theory of dynamical systems—how they should be approximated becomes a pressing question. Also measures provide models for greytone photographs: their approximation is equivalent to image compression!

What sort of things will Fractal Approximation Theory use as approximating entities? And how are the approximating entities to be computed? In this issue the focus is on the use of deterministic fractals and invariant measures of Markov processes as approximating entities. The papers by Bedford, and by Barnsley, Elton, and Hardin are in this area. One approach is the use of attractors of iterated function systems built up from a special class of transformations in the underlying space, such as affine transformations. The approximants introduced by Elton and Yan give the flavor of this type of problem. They introduce a special class of approximants which are quite easy to work with.

The approximating entities are computed by set iteration, and by random iteration algorithms which exploit ergodic theorems, such as the ergodic theorem of Elton. Withers uses Newton's method to compute a sequence of fractal approximations to the graph of a given function, and he uses ergodicity to compute the necessary derivative at each step!

Third, what criteria will Fractal Approximation Theory use to decide whether or not an approximation is a good one? What quantities do we try to make agree between the approximations and the objects which are being approximated? What are the distances between target and approximation which are to be minimized? The Hausdorff dimension of approximating sets associated with iterated function systems of affine maps is dealt with in the paper of Geronimo and Hardin. The paper of Bedford introduces the notion of the distribution of Hölder exponents as a characteristic of the graph of a nondifferentiable function. He gives explicit formulas relating these important approximation criteria numbers to fractal dimensions.

The paper by Wallin is important because it underlies another fascinating area: the approximation and continuation properties whose domains are fractal subsets of $\mathbf{R}^n$. How does one go about approximating the temperature of the coast of Sweden?

Michael Barnsley
Georgia Institute of Technology
Atlanta, Georgia

Constr. Approx. (1989) 5: 3–31

CONSTRUCTIVE APPROXIMATION

Recurrent Iterated Function Systems

Michael F. Barnsley, John H. Elton, and Douglas P. Hardin

Abstract. Recurrent iterated function systems generalize iterated function systems as introduced by Barnsley and Demko [BD] in that a Markov chain (typically with some zeros in the transition probability matrix) is used to drive a system of maps $w_j: K \to K$, $j = 1, 2, \ldots, N$, where K is a complete metric space. It is proved that under "average contractivity," a convergence and ergodic theorem obtains, which extends the results of Barnsley and Elton [BE]. It is also proved that a Collage Theorem is true, which generalizes the main result of Barnsley *et al.* [BEHL] and which broadens the class of images which can be encoded using iterated map techniques. The theory of fractal interpolation functions [B] is extended, and the fractal dimensions for certain attractors is derived, extending the technique of Hardin and Massopust [HM]. Applications to Julia set theory and to the study of the boundary of IFS attractors are presented.

1. Introduction

Let X be a complete metric space with metric d. Let $w_j: X \to X$ be Lipschitz maps, $j = 1, 2, \ldots, N$. Let (p_{ij}) be an $N \times N$ row-stochastic matrix. Then we call $\{X, w_j, p_{ij}, i, j = 1, 2, \ldots, N\}$ a recurrent iterated function system (IFS)—whether or not (p_{ij}) is technically "recurrent" (i.e., irreducible). The focus of a recurrent IFS is random walks in X of the following nature: specify a starting point $x_0 \in X$ and a starting code $i_0 \in \{1, 2, \ldots, N\}$. Choose a number $i_1 \in \{1, 2, \ldots, N\}$ with the (conditional) probability that $i_1 = j$ being $p_{i_0 j}$, and then define $x_1 = w_{i_1} x_0$. Then pick $i_2 \in \{1, 2, \ldots, N\}$, with the probability that $i_2 = j$ being $p_{i_1 j}$, and go to the point $x_2 = w_{i_2} x_1 = w_{i_2} w_{i_1} x_0$. Continue in this way to generate an orbit $\{x_n\}_{n=0}^{\infty}$.

Our concern in this paper is with existence, uniqueness, convergence to, and characterization of limit sets (attractors) $A \subset X$, and of associated invariant (stationary) measures whose support is A. A may be described as follows: $x \in A$ iff every neighborhood of x contains infinitely many x_n's, for almost all orbits. The empirical distribution along an orbit converges to the stationary measure, for almost all orbits. (The description given of A does not quite follow from the statement about the stationary measure, which is of interest; see Section 2.)

It is also very important to consider limits when composing maps in the reverse order $w_{i_1} \cdots w_{i_n} x$, which is exploited, and connections with the random walk

Date received: September 4, 1986. Communicated by Edward B. Saff.
AMS classification: 28D99, 41A99, 58F11, 60F05, 60G10, 60J05.
Key words and phrases: Iterated function systems, Attractor, Random maps, Markov chain, Ergodic, Lyapunov exponent, Fractal, Dimension.

clarified, in Section 2. In the case of uniformly contractive maps this is especially useful and in Section 3 this point of view is exclusively used, in a more general setting, to give an elegant characterization of the attractor as the unique, attractive fixed point of a certain set map, using the Hausdorff metric. (Actually, a more precise invariance result is obtained for an N-tuple of sets, based on the connection structure of the chain—that is, which maps are allowed to follow which, i.e., which p_{ij} are not zero.)

By having some entries in (p_{ij}) equal to zero, the allowable map sequences in the random walk are restricted, and this gives rise to limit sets with geometries not obtainable by earlier iterated function systems, which is one motivation for our work; see Section 5.

Key references which underlie the present work are [BD], [H], [Bed], [D], [BEHL], [HM], [BE], [E], and [BA]. The structure of this paper is as follows. In Section 2 we consider the existence, uniqueness, and convergence questions referred to above. In Section 3 we describe the College Theorem for recurrent IFS, and in so doing extend the concept of recurrent IFS to multiple spaces and set maps. In Section 4 we compute the fractal dimension for various recurrent IFS attractors, using the Perron–Frobenius theorem for the connection matrix. In Section 5 we give examples, including combinatorial fractal functions, boundaries of attractors of IFS, and Julia set applications.

2. Ergodicity of the Random Walk

Let (X, d) be a complete separable locally compact metric space. We consider a random walk (i.e., a discrete-time stochastic process) in X arising from iteratively applying Lipschitz maps chosen according to a finite state-space Markov chain, as described in the Introduction.

Let (p_{ij}) be an irreducible $N \times N$ row-stochastic matrix, i.e., $\sum_{j=1}^{N} p_{ij} = 1$ for all i, $p_{ij} \geq 0$ for all i, j, and for any i, j there exist $i_1, i_2, \ldots, i_n$ with $i_1 = i$ and $i_n = j$ such that $p_{i_1 i_2} p_{i_2 i_3} \cdots p_{i_{n-1} i_n} > 0$. Let w_j, $j = 1, \ldots, N$, be Lipschitz maps on X.

The random walk described informally in the Introduction can be formulated as follows: let $i_0, i_1, \ldots$ be a Markov chain in $\{1, \ldots, N\}$, with transition probability matrix (p_{ij}); then our random walk is the process

$$Z_n = w_{i_n} Z_{n-1} = w_{i_n} \cdots w_{i_1} Z_0.$$

Now (Z_n) is not a Markov process on X, but $\tilde{Z}_n = (Z_n, i_n)$ is a Markov process on $\tilde{X} = X \times \{1, \ldots, N\}$ with transition probability function

$$\tilde{p}((x, i), \tilde{B}) = \sum_{j=1}^{N} p_{ij} I_{\tilde{B}}(w_j x, j);$$

this is the probability of transfer from (x, i) into the Borel set $\tilde{B} \subset \tilde{X}$ in one step of the process.

Let (m_i) be the unique stationary initial distribution for the Markov chain on $\{1, \ldots, N\}$; i.e.,

$$\sum_{i=1}^{N} m_i p_{ij} = m_j, \qquad j = 1, \ldots, N.$$

We show that if the maps are logarithmically contractive on the average after some number of iterations (see Theorem 2.1), then there is a unique initial distribution which makes the Markov process $(\tilde{Z}_n)$ stationary (this is also called the invariant measure), and more importantly, for *any* starting value (x_0, i_0), the empirical distribution of a trajectory $x_0, w_{i_1}x_0, w_{i_2}w_{i_1}x_0, \ldots$ will converge with probability one to the X-projection μ of the stationary initial distribution. Furthermore, if A is the support of μ, then $x \in A$ iff for any neighborhood of x, almost all trajectories visit the neighborhood infinitely often. This is perhaps surprising because from the convergence to μ of the empirical distribution along trajectories it follows that $x \notin A \Rightarrow$ for some neighborhood of x, the proportion of the number of visits to the neighborhood approaches 0 for almost all trajectories, whereas we are making the stronger assertion that some neighborhood of x will only be visited finitely many times, almost surely (a.s.).

This average contractivity condition appeared in [BE] concerning the case when the sequence of maps is independent and identically distributed (i.i.d.) (in the present setup, this would mean $p_{ij} = p_j$, $i = 1, \ldots, N$, for each j), and was generalized in the case of i.i.d. affine maps to infinitely many maps in [BA]. It is equivalent in those cases to a negative Lyapunov exponent condition, as pointed out in [BA], and this will be seen to be true here also.

An important point in the proof will be to run the Markov chain "backward in time"; let us explain. Consider the matrix

$$q_{ij} = \frac{m_j}{m_i} p_{ji},$$

which is row stochastic, irreducible, and also satisfies $\sum_{i=1}^{N} m_i q_{ij} = m_j$, as is easily checked. The q_{ij} are called *inverse* transition probabilities in Chapter 15 of [F]. The reason is as follows: consider the Markov chain $(i_0, i_1, \ldots)$ above, with transition probability matrix (p_{ij}) and initial distribution (m_i). The probability that $(i_1, i_2, \ldots, i_n) = (j_1, \ldots, j_n)$ is then

$$\sum_{j_0=1}^{N} m_{j_0} p_{j_0 j_1} p_{j_1 j_2} \cdots p_{j_{n-1} j_n} = \sum_{j_0=1}^{N} m_{j_0} \frac{m_{j_1}}{m_{j_0}} q_{j_1 j_0} \cdots \frac{m_{j_n}}{m_{j_{n-1}}} q_{j_n j_{n-1}}$$

$$= m_{j_n} q_{j_n j_{n-1}} \cdots q_{j_2 j_1},$$

which is the probability that a Markov chain with transition probability matrix (q_{ij}) and initial distribution (m_i) will have for its first n values $(j_n, j_{n-1} \cdots j_1)$.

Let P be the probability measure on $\Omega = \{\mathbf{i} = (i_0, i_1, \ldots)\}$ corresponding to the "forward" chain; that is, P is given on "thin cylinders" by $P(i_0, i_1, \ldots, i_n) = m_{i_0} p_{i_0 i_1} \cdots p_{i_{n-1} i_n}$. Let Q be the probability measure on Ω corresponding to the "backward" chain; i.e., $Q(i_0, i_1, \ldots, i_n) = m_{i_0} q_{i_0 i_1} \cdots q_{i_{n-1} i_n}$.

We show that under our hypotheses, for the backward process, $\lim_{n \to \infty} w_{i_1} \cdots w_{i_n} x = Y(\mathbf{i})$ exists and is independent of x for Q—almost all (a.a.) $\mathbf{i}$. Note that this is *very different* from the iterative process $w_{i_n} \cdots w_{i_1} x$ we originally discussed, where $i_1, i_2, \ldots$ are chosen according to P. This process does *not* converge pointwise, but its trajectories distribute ergodically as the measure μ which is obtainable from the limit of the backward process as

$\mu(B) = Q(Y^{-1}(B))$. This is simply because for all n, $w_{i_n} \cdots w_{i_1}x$ has the same *distribution* under P as does $w_{i_1} \cdots w_{i_n}x$ under Q.

If all the maps w_j are *uniform contractions*, then it is easy to see that $\lim_{n\to\infty} w_{i_1} \cdots w_{i_n}x = Y(\mathbf{i})$ exists for *all* $\mathbf{i}$ (not just Q a.e.), and that Y is continuous with the product topology on Ω, and range $Y = A$ is a compact set in X which is exactly the support of μ, called the *attractor*. This is discussed in detail in Section 3, where an invariance result ("Collage Theorem") is given for a special decomposition of A into compact subsets. But even in this uniformly contractive case, the trajectories of the random walk (the forward process), $w_{i_n} \cdots w_{i_1}x$, converge only in the distribution sense (the points along the trajectory continue to dance about), and only with probability one.

We hope this detailed discussion of running time backward and inverse probabilities will be helpful in clarifying the connection between the "symbolic dynamics" point of view $w_{i_1}w_{i_2} \cdots w_{i_n}x$ and the "ergodic" point of view $w_{i_n} \cdots w_{i_1}x$, and why the measures are the same; this matter had been a little unclear to the authors previously.

Now for a precise statement and proof of the convergence and ergodic results. For a Lipschitz map $w: X \to X$, define

$$\|w\| = \sup_{x\neq y} \frac{d(wx, wy)}{d(x, y)}.$$

Theorem 2.1. *Assume that, for some n,*

$$\mathbf{E}_P(\log\|w_{i_n} \cdots w_{i_1}\|) < 0$$

(see above for the definition of P); that is,

$$\sum_{i_i} \cdots \sum_{i_n} m_{i_1} p_{i_1 i_2} \cdots p_{i_{n-1} i_n} \log\|w_{i_1} \cdots w_{i_n}\| < 0.$$

(This is equivalent to a negative Lyapunov exponent for the process $w_{i_1}, w_{i_2}, \ldots$; see the proof). Then:

(i) *For Q a.a. $\mathbf{i}$, $w_{i_1} \cdots w_{i_n}x \to Y(\mathbf{i})$, which does not depend on the choice of $x \in X$.*

(ii) *Define $\tilde{\mu}(\tilde{B}) = Q(\mathbf{i}\colon (Y(\mathbf{i}), i_1(\mathbf{i})) \in \tilde{B})$, the distribution of (Y, i_1) on $\tilde{X}$. Then $\tilde{\mu}$ is the unique stationary initial distribution for the Markov process $\tilde{Z}_n = (Z_n, i_n)$. Furthermore, if $\tilde{\nu}$ is any probability measure on $\tilde{X}$ satisfying $\tilde{\nu}(X\times\{i\}) = m_i$, $i = 1, \ldots, N$, then $\tilde{Z}_n^{\tilde{\nu}}$ converges in distribution to $\tilde{\mu}$, where $\tilde{Z}_n^{\tilde{\nu}}$ represents the Markov process with initial distribution $\tilde{\nu}$. In particular, the random walk $Z_n^{\tilde{\nu}}$ on X converges in distribution to the measure $\mu(B) = \tilde{\mu}(B\times\{1, \ldots, N\})$. (The given condition on $\tilde{\nu}$ may be expressed as requiring the marginal distribution of $i_0^{\tilde{\nu}}$ to be (m_i).)*

(iii) *(Ergodic theorem). For* every *x, for P a.a. $\mathbf{i}$,*

$$\frac{1}{n}\sum_{k=1}^{n} f(w_{i_k} \cdots w_{i_1}x) \to \int f\,d\mu$$

for all $f \in C(X)$, the bounded continuous functions on X. In other words, starting at any *x, the empirical distribution of a trajectory converges with probability one to μ.*

(iv) *If A is the support of μ, then $x \in A$ iff for every neighborhood of x, almost all trajectories visit the neighborhood infinitely often (recall that the* support *A of μ is defined as follows: $x \in A$ iff every neighborhood of x has positive μ-measure; A is a closed set).*

Proof. (i) First note that since the distribution of $(i_1 \cdots i_n)$ under P is the same as $(i_n, \ldots, i_1)$ under Q as pointed out above, we have $\mathbf{E}_Q \log\|w_{i_1} \cdots w_{i_n}\| < 0$ also. Since $(i_1 i_2, \ldots)$ is a stationary ergodic process under P and Q, the proof of the Furstenberg-Kesten theorem given on p. 40 of [Kr] shows that this is equivalent to

$$\lim_{n\to\infty} \frac{1}{n} \log\|w_{i_n} \cdots w_{i_1}\| = -\alpha, \quad P \text{ a.s.},$$

and to

$$\lim_{n\to\infty} \frac{1}{n} \log\|w_{i_1} \cdots w_{i_n}\| = -\alpha, \quad Q \text{ a.s.},$$

where $\alpha > 0$ ($-\alpha$ is the Lyapunov exponent). (The proof in [Kr] refers to linear maps, but there is no change in the proof needed for our case, or for reversing the order.)

For the remainder of the proof of (i), we borrow the more elegant method of [BA], rather than the earlier proof of [BE]. Fix x. Now

$$d(w_{i_1} \cdots w_{i_n} x, w_{i_1} \cdots w_{i_{n+1}} x) \le \|w_{i_1} \cdots w_{i_n}\| C(x),$$

where $C(x) = \max_{1\le i\le N} d(w_i x, x)$. For Q a.a. $\mathbf{i}$, we may choose n_0 (depending on $\mathbf{i}$) so that $n \ge n_0 \Rightarrow \|w_{i_1} \cdots w_{i_n}\| < e^{-n\alpha/2}$. Thus

$$\textstyle\sum_{n=1}^{\infty} d(w_{i_1} \cdots w_{i_n} x, w_{i_1} \cdots w_{i_{n+1}} x) < \infty,$$

so $w_{i_1} \cdots w_{i_n} x$ is Cauchy and converges to say $Y(\mathbf{i})$, for Q a.a. $\mathbf{i}$. Furthermore, $d(w_{i_1} \cdots w_{i_n} x, w_{i_1} \cdots w_{i_n} y) \le \|w_{i_1} \cdots w_{i_n}\|\, d(x, y) \to 0$ Q a.s., so Y does not depend on x.

(ii) Let $\tilde{\nu}$ be any probability measure on $\tilde{X}$ satisfying $\tilde{\nu}(X \times \{i\}) = m_i$, $i = 1, \ldots, N$. Then if the Markov process is given initial distribution $\tilde{\nu}$, $\mathbf{i}$ will have distribution P, since i_0 will have marginal distribution (m_i). For each j, let

$$\nu_j(B) = \frac{\tilde{\nu}(B \times \{j\})}{\tilde{\nu}(X \times \{j\})}$$

(note $\tilde{\nu}(X \times \{j\}) = m_j > 0$ for all j). This is the conditional distribution of Z_0 given $i_0 = j$. Thus for all $\tilde{f} \in C(\tilde{X})$, $\mathbf{E}\tilde{f}(\tilde{Z}_n^{\tilde{\nu}}) = \iint \tilde{f}(w_{i_n} \cdots w_{i_1} x, i_n)\, d\nu_{i_0}(x)\, dP(\mathbf{i}) = \iint \tilde{f}(w_{i_1} \cdots w_{i_n} x, i_1)\, d\nu_{i_{n+1}}(x)\, dQ(\mathbf{i})$. Now fix x_0. For Q a.a. $\mathbf{i}$, $d(w_{i_1} \cdots w_{i_n} x, w_{i_1} \cdots w_{i_n} x_0) \to 0$ for *every* x, so

$$\iint \tilde{f}(w_{i_1} \cdots w_{i_n} x, i_1)\, d\nu_{i_{n+1}}(x)\, dQ(\mathbf{i}) - \iint \tilde{f}(w_{i_1} \cdots w_{i_n} x_0, i_1)\, d\nu_{i_{n+1}}(x)\, dQ(\mathbf{i}) \to 0.$$

But

$$\int\int \tilde{f}(w_{i_1}\cdots w_{i_n}x_0, i_1)\, d\nu_{i_{n+1}}(x)\, dQ(\mathbf{i})$$

$$= \int \tilde{f}(w_{i_1}\cdots w_{i_n}x_0, i_1)\, dQ(\mathbf{i}) \to \int \tilde{f}(Y, i_1)\, dQ = \int \tilde{f}\, d\tilde{\mu}.$$

This shows $\tilde{Z}_n^{\tilde{\nu}}$ converges in distribution to $\tilde{\mu}$.

It remains to show that $\tilde{\mu}$ is a stationary initial distribution, and is unique.

Let $\tilde{\nu}$ be *any* stationary initial distribution. Then $\tilde{\nu}(X \times \{i\}) = m_i$, $i = 1, \ldots, N$, since i_0 must have marginal distribution (m_i) in order that (i_n) be stationary since the chain is irreducible. For $\tilde{f} \in C(\tilde{X})$, let $T\tilde{f}(\tilde{x}) = \mathbf{E}\tilde{f}(\tilde{Z}_1^{\tilde{x}})$, where $\tilde{Z}_n^{\tilde{x}}$ is the Markov process with $\tilde{Z}_0^{\tilde{x}} \equiv \tilde{x}$; this is the usual Markov operator on $C(\tilde{X})$. The adjoint T^* restricted to Borel measures has the following interpretation: if $\tilde{\nu}$ is the distribution of $\tilde{Z}_0$, then $T^*\tilde{\nu}$ is the distribution of $\tilde{Z}_1$, and $T^{*n}\tilde{\nu}$ is the distribution of $\tilde{Z}_n$. Thus from what was just shown, if $\tilde{\nu}$ is a stationary initial distribution, $\tilde{\nu} = T^{*n}\tilde{\nu} \to^{W^*} \tilde{\mu}$, so $\tilde{\mu}$ is the only possible stationary initial distribution. Furthermore, if $\tilde{\nu}$ is *any* distribution satisfying $\tilde{\nu}(X \times \{i\}) = m_i$ for all i, then $T^*\tilde{\nu}$ satisfies this condition also, since $T^*\tilde{\nu}(X \times \{i\})$ is just the marginal distribution of $i_1^{\tilde{\nu}}$, which is (m_i) since $i_0^{\tilde{\nu}}$ was given the stationary initial distribution (m_i). Thus choosing $\tilde{\nu}$ to be, for example, $\delta_x \times (m_i)$, we have $T^{*n}\tilde{\nu} \to^{W^*} \tilde{\mu}$, and $T^*(T^{*n}\tilde{\nu}) \to^{W^*} T^*\tilde{\mu}$ since T^* is $w^* - w^*$ continuous; but $T^*(T^{*n}\tilde{\nu}) = T^{*n}(T^*\tilde{\nu}) \to^{W^*} \tilde{\mu}$ also, so $T^*\tilde{\mu} = \tilde{\mu}$, so $\tilde{\mu}$ is stationary.

(iii) This is the only place where the assumption that X is separable and locally compact is needed.

Now $\tilde{Z}_n^{\tilde{\mu}}$ is an *ergodic* stationary process since $\tilde{\mu}$ is the unique stationary initial distribution (see Lemma 1 of [E]). Now let $f \in C_c(X)$, the continuous functions on X with compact support. By the classical pointwise ergodic theorem, with probability one,

$$\frac{1}{n}\sum_{k=1}^{n} f(Z_k^{\tilde{\mu}}) \to \int f\, d\mu$$

(we consider f to be defined on $\tilde{X}$ by $f(x, i) = f(x)$). But $d(Z_k^{\tilde{\mu}}, Z_k^x) = d(w_{i_n}\cdots w_{i_1}Z_0, w_{i_n}\cdots w_{i_1}x) \le \|(w_{i_n}\cdots w_{i_1}\|\, d(Z_0, x) \to 0$ with probability one, where $\tilde{Z}_k^x$ is the Markov process with $Z_0^x \equiv x$ and i_0^x distributed as (m_i). Since f is *uniformly* continuous, we get

$$\frac{1}{n}\sum_{k=1}^{n} f(Z_k^x) = \frac{1}{n}\sum_{k=1}^{n} f(w_{i_k}\cdots w_{i_1}x) \to \int f\, d\mu$$

also, with probability one. Since $C_c(X)$ is separable, we can get this for all $f \in C_c(X)$ *simultaneously*, with probability one. Finally, a simple argument using Urysohn's lemma extends this to $C(X)$ (note (Z_n^x) is tight since it converges in distribution).

(iv) $x \in A \Rightarrow$ for any neighborhood of x, almost all trajectories visit the neighborhood infinitely often follows immediately from (iii), which says that in fact that the proportion of visits is asymptotically positive.

Going in the other direction, assume $x \notin A$. Let $x_0 \in X$. We want to show that there is some neighborhood of x such that almost all trajectories starting at x_0 visit the neighborhood only finitely often. Since $x \notin A$, $d(x, A) = \varepsilon > 0$. For P a.a. **i**, there is n_0 (depending on **i**) such that $n > n_0 \Rightarrow \|w_{i_n} \cdots w_{i_1}\| < \varepsilon/(4(d(y, x_0)+1))$, from the proof of (i). Also, for P a.a. **i**, $P((i_1, \ldots, i_n)) > 0$ for all n (that is, $p_{i_1 i_2} p_{i_2 i_3} \cdots p_{i_{n-1} i_n} > 0$), so $Q((i_n, \ldots, i_1)) > 0$ also. So fix $i_1 \cdots i_n$ so that $\|w_{i_n} \cdots w_{i_1}\| > \varepsilon/(4(d(x_0, y)+1))$ and $P((i_1, \ldots, i_n)) > 0$. Let $y \in A$. Now $Y \in A$ Q a.s. by definition of A, so for Q a.a. $(j_1, j_2, \ldots)$ we have $\lim_{k\to\infty} w_{i_n} \cdots w_{i_1} w_{j_1} \cdots w_{j_k} y \in A$. Also from the definition of A, the set $\{\lim w_{j_1} \cdots w_{j_k} y : \mathbf{j} \in J\}$ is dense in A for any J with $Q(J) = 1$. Thus we may find **j** and k such that $d(w_{j_1} \cdots w_{j_k} y, y) < 1$ and $d(w_{i_n} \cdots w_{i_1} w_{j_1} \cdots w_{j_k} y, A) < \varepsilon/4$. Now

$$d(w_{i_n} \cdots w_{i_1} y, A) \leq d(w_{i_n} \cdots w_{i_1} y, w_{i_n} \cdots w_{i_1} w_{j_1} \cdots w_{j_k} y)$$
$$+ d(w_{i_n} \cdots w_{i_1} w_{j_1} \cdots w_{j_k} y, A) \leq \|w_{i_n} \cdots w_{i_1}\| \, d(y, w_{j_1} \cdots w_{j_k} y) + \varepsilon/4 < \varepsilon/2.$$

Thus

$$d(w_{i_n} \cdots w_{i_1} x_0, A) \leq d(w_{i_n} \cdots w_{i_1} x_0, w_{i_n} \cdots w_{i_1} y) + d(w_{i_n} \cdots w_{i_1} y, A)$$
$$\leq \|w_{i_n} \cdots w_{i_1}\| \, d(x_0, y) + \varepsilon/2 < 3\varepsilon/4.$$

Thus $w_{i_n} \cdots w_{i_1} x_0$ is not in the ball of radius $\varepsilon/4$ centered at x. ■

3. Point Set Topology: The Collage Theorem for Recurrent IFS

3.1. Hyperbolic Recurrent IFS

We recall that a recurrent IFS is described in terms of a Markov chain which acts on a code space built from the symbols $\{1, 2, \ldots, N\}$. Such a process with $N = 3$ may be described by a directed graph: see, for example, Fig. 1 where the numbers $p_{ij} \geq 0$, $\sum_{j=1}^{N} p_{ij} = 1$, give the probabilities of transfer among the symbols. We can image a particle moving from symbol to symbol following the discrete-time

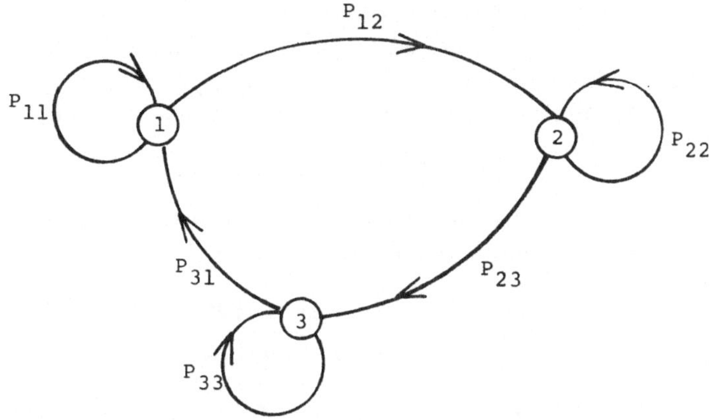

Fig. 1

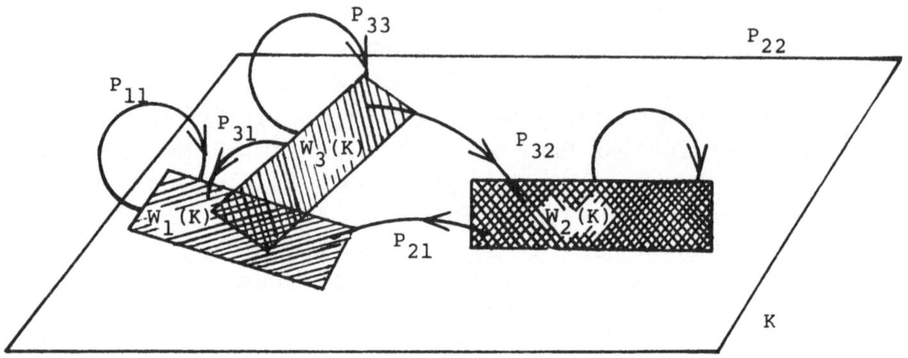

Fig. 2

Markov process. The process is defined, strictly speaking, to be *recurrent* if there is a finite probability of being able to move, on the directed graph, from any given symbol to any given symbol. A good source of information on Markov chains is Chapter 15 of [F].

The central idea of recurrent IFS theory is that such a Markov chain is used to drive the application of maps $w_i: K \to K$, $i = 1, 2, \ldots, N$, where K, for the purposes of this section, is a compact metric space for simplicity. Such a process is symbolized in Fig. 2. (In distinction to what happens on the symbols, the fact that we have just applied map w_1, for example, does not mean we are in the same place as we were the last time map w_1 had been applied: the sequence of points $\{x_n\}_{n=0}^{\infty}$ contains many more than three values!)

We are here concerned with the *hyperbolic* case, namely,

$$d(w_i(x), w_i(y)) \le sd(x, y), \qquad \forall i, \forall x, y \in K,$$

where d is the distance function on K and $0 \le s < 1$. In this case, from Section 2, we know that there exists a unique attractive invariant probability measure μ, which describes the random walk on K. Our focus in this section is on the structure of the *support* of μ, which we call $A \subset K$, the attractor of the recurrent IFS. This depends only on which p_{ij} are nonzero, the *connection structure* of the chain, and otherwise not on the values of the p_{ij}'s.

Note that IFS theory, as described in [BD], corresponds to Markov chains whose transition matrices have the special structure $p_{ij} = p_j > 0$, $i, j = 1, 2, \ldots, N$, as symbolized in the directed graph shown in Fig. 3 which corresponds to $N = 3$.

3.2. Hausdorff Distance

Let (K, d) denote a compact metric space K, with distance function d. Let H denote the set of all nonempty compact subsets of K.

Definition 3.1. $d(x, B) = \min_{y \in B} d(x, y)$, $\forall x \in K$, $\forall B \in H$. Note that

$$(*) \qquad B \subset C \;\Rightarrow\; d(x, C) \le d(x, B).$$

Definition 3.2. $d(A, B) = \max_{x \in A} d(x, B)$, $\forall A, B \in H$. Note that this "distance" is not symmetric.

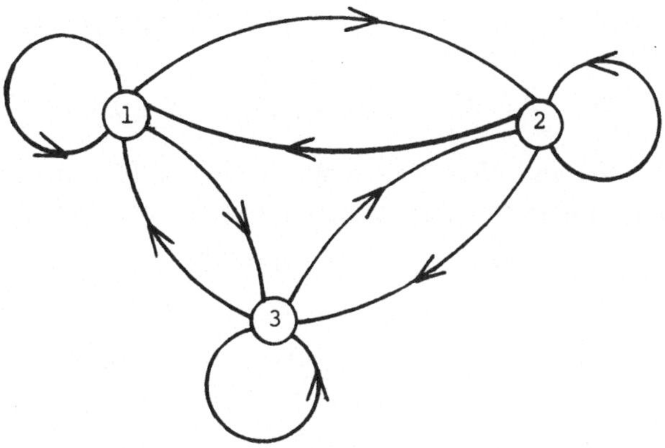

Fig. 3

Properties.

(i) $B \subset C \Rightarrow d(A, C) \leq d(A, B)$, by (*).

(ii) $d(A \cup B, C) = d(A, C) \vee d(B, C)$, where $x \vee y = \max\{x, y\}$,

$$d(A \cup B, C) = \max_{x \in A \cup B} d(x, C) = \max_{x \in A} d(x; C) \vee \max_{x \in B} d(x, C).$$

Definition 3.3. For all A, $B \in H$, the Hausdorff distance is defined by

$$h(A, B) = d(A, B) \vee d(B, A).$$

Remark. (H, h) is a compact metric space [Dj].

Lemma 3.1. *For all* A, B, C, $D \in H$,

$$h(A \cup B, C \cup D) \leq h(A, C) \vee h(B, D).$$

Proof.

$$\begin{aligned} d(A \cup B, C \cup D) &= d(A, C \cup D) \vee d(B, C \cup D) \quad \text{by (ii)} \\ &\leq d(A, C) \vee d(B, D) \quad \text{by (i)} \\ &\leq h(A, C) \vee h(B, D). \end{aligned}$$

The same argument yields

$$d(C \cup D, A \cup B) \leq d(C, A) \vee d(B, D) \leq h(A, C) \vee h(B, D). \qquad \blacksquare$$

3.3. *The Standard Collage Theorem* [BEHL]

Let $\{K, w_j, j = 1, 2, \ldots, N\}$ be a hyperbolic IFS, with $d(w_j x, w_j y) \leq s d(x, y)$, $\forall x, y \in K$, and $0 \leq s < 1$. Define

$$W: H \to H$$

by

$$W(A) = \bigcup_{j=1}^{N} w_j(A).$$

Theorem 3.1. *$W: H \to H$ is a contraction, with Lipschitz constant s, with respect to the Hausdorff metric; that is,*

$$h(W(A), W(B)) \leq sh(A, B), \qquad \forall A, B \in H.$$

Remark. If (K_1, d_1) and (K_2, d_2) are metric spaces, (H_1, h_1) and (H_2, h_2) are the corresponding spaces of compact nonempty subsets, and if

$$w_j: K_1 \to K_2 \qquad \text{for} \quad j = 1, 2, \ldots, N$$

obeys

$$d_2(w_j x, w_j y) \leq sd_1(x, y), \qquad \forall x, y \in K_1,$$

then

$$W: H_1 \to H_2$$

defined by

$$W(A) = \bigcup_{j=1}^{N} w_j(A)$$

obeys

$$h_2(W(A), W(B)) \leq sh_1(A, B).$$

We will need this extension of Theorem 3.1 later.

Corollary 3.2. *There is a unique set $A \in H$ such that $W(A) = A$.*

Remark. A is the attractor of the IFS. It is the support of any p-balanced measure associated with the IFS [BD].

Corollary 3.3 (Collage Theorem [BEHL]). *If $B \in H$ obeys*

$$h(B, W(B)) \leq \varepsilon > 0,$$

then

$$h(B, A) \leq \varepsilon/(1-s),$$

where A denotes the attractor of the IFS.

Proof of Theorem 3.1. For any $A, B \in H$,

$$\begin{aligned} h(W(A), W(B)) &= h\left(\bigcup_{j=1}^{N} w_j(A), \bigcup_{j=1}^{N} w_j(B)\right) \\ &\leq \bigvee_{j=1}^{N} h(w_j(A), w_j(B)) \quad \text{(by Lemma 3.1)} \\ &= \bigvee_{j=1}^{N} \{d(w_j(A), w_j(B)) \vee d(w_j(B), w_j(A))\} \\ &\leq \bigvee_{j=1}^{N} \{sd(A, B) \vee sd(B, A)\} \\ &= sh(A, B). \end{aligned}$$

■

Proof of Corollary 3.3. By the contraction mapping theorem

$$h(A, B) = h(B, \lim_{n\to\infty} W^{\circ n}(B)) = \lim_{n\to\infty} h(B, W^{\circ n}(B)),$$

where $W^{\circ 0}(B) = W(B)$ and we define inductively

$$W^{\circ(n+1)}(B) = W(W^{\circ n}(B)), \qquad n = 0, 1, 2, \ldots.$$

But by the triangle inequality

$$\begin{aligned} h(B, W^{\circ n}(B)) &\leq \sum_{m=1}^{n} h(W^{\circ(m-1)}(B), W^{\circ m}(B)) \\ &= \sum_{m=1}^{n} h(W^{\circ(m-1)}(B), W^{\circ(m-1)}(W(B))) \\ &\leq \sum_{m=1}^{n} s^{m-1} h(B, W(B)) \\ &\leq (1-s)^{-1} h(B, W(B)). \end{aligned}$$

■

3.4. The Collage Theorem for Recurrent IFS

We actually make a generalization of the recurrent IFS structure to multiple spaces and set maps, suitable for the hyperbolic case where we are concerned with point-set topology issues. We are only concerned here with the connection structure of the chain. Let (K_j, d_j) be compact metric spaces, $j \in \{1, 2, \ldots, N\}$. Let (H_j, h_j) denote the associated metric spaces of nonempty compact subsets which use the Hausdorff metrics. Let there be defined maps $W_{ij}: H_j \to H_i, \forall (i, j) \in I$, where I is some set of pairs of indices with the property that for each $i \in \{1, 2, \ldots, N\}$ there is a $j \in \{1, 2, \ldots, N\}$ with $(i, j) \in I$. That is, $I(i) = \{j \mid (i, j) \in I\} \neq \varnothing$ for each $i \in \{1, 2, \ldots, N\}$. Furthermore, let

$$h_i(W_{ij}(A), W_{ij}(B)) \leq s_{ij} h_j(A, B)$$

for some number s_{ij}, $\forall (i, j) \in I$, $\forall A, B \in H_j$. By the remark following Theorem 3.1 such maps can be built up from point maps taking K_j to K_i.

The setup of the Introduction, Section 2, and Subsection 3.1 can be put into this more general framework as follows: assume $w_i: K \to K$ are Lipschitz maps, where K is compact metric, and (p_{ij}) is row stochastic. Define $(K_j, d_j) = (K, d)$ for each j, and define $W_{ij}(S) = \{w_i(x): x \in S\}$, $i, j = 1, \ldots, N$. Let $I(i) = \{j: p_{ji} > 0\}$. This embeds us in the more general setup which we now study.

Let

$$\tilde{H} = H_1 \times H_2 \times H_3 \times \cdots \times H_N,$$

and endow $\tilde{H}$ with the metric $\tilde{h}$ defined by

$$\tilde{h}((A_1, A_2, \ldots, A_N), (B_1, B_2, \ldots, B_N)) = \max\{h_j(A_j, B_j) \mid j = 1, 2, \ldots, N\}.$$

Then it is readily demonstrated that $(\tilde{H}, \tilde{h})$ is a compact metric space.

We think of $\tilde{H}$ as consisting of a stack of clipped planes $K_1, K_2, \ldots, K_N$ with

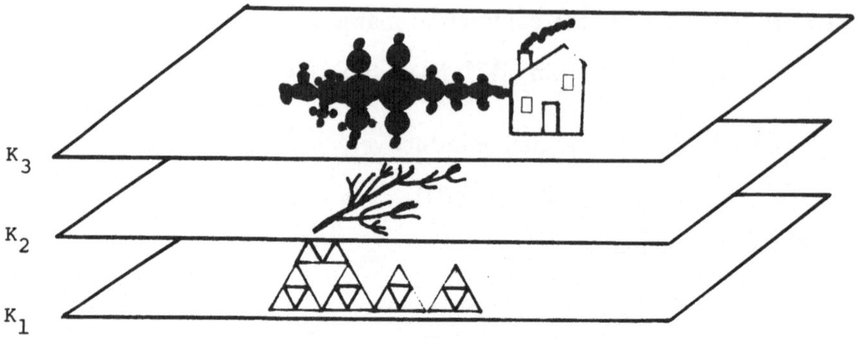

Fig. 4

a point in $\tilde{H}$ being the N-tuple of one image in each plane (see Fig. 4). Define

$$W: \tilde{H} \to \tilde{H}$$

by

$$W(A_1, A_2, \ldots, A_N) = \left(\bigcup_{j\in I(1)} w_{1j}(A_j), \bigcup_{j\in I(2)} w_{2j}(A_j), \ldots, \bigcup_{j\in I(N)} w_{Nj}(A_j) \right).$$

Example. Such a mapping with $N=2$ might be symbolized

$$W\begin{pmatrix} A_1 \\ A_2 \end{pmatrix} = \begin{pmatrix} \varnothing & W_{12} \\ W_{21} & W_{22} \end{pmatrix} \begin{pmatrix} A_1 \\ A_2 \end{pmatrix} = \begin{pmatrix} W_{12}(A_2) \\ W_{21}(A_1) \cup W_{22}(A_2) \end{pmatrix}.$$

Theorem 3.4. *$W: \tilde{H} \to \tilde{H}$ obeys*

$$\tilde{h}(W(A), W(B)) \le s\tilde{h}(A, B), \qquad \forall A, B \in H,$$

where $s = \max\{s_{ij}, (i,j) \in I\}$.

Proof. To keep the notation succinct we assume

$$W = \begin{pmatrix} W_{11} & W_{12} \\ W_{21} & W_{22} \end{pmatrix}.$$

Then if $A = (A_1, A_2)$ and $B = (B_1, B_2)$ we have

$$\begin{aligned} \tilde{h}(W(A), W(B)) &= \tilde{h}((W_{11}(A_1) \cup W_{12}(A_2), W_{21}(A_1) \cup W_{22}(A_2)), \\ &\qquad (W_{11}(B_1) \cup W_{12}(B_2), W_{21}(B_1) \cup W_{22}(B_2))) \\ &= \max\{h_1(W_{11}(A_1) \cup W_{12}(A_2), W_{11}(B_1) \cup W_{12}(B_2)), \\ &\qquad h_2(W_{21}(A_1) \cup W_{22}(A_2), W_{21}(B_1) \cup W_{22}(B_2))\} \\ &\le \max\{h_1(W_{11}(A_1), W_{11}(B_1)) \vee h_1(W_{12}(A_2), W_{12}(B_2)), \\ &\qquad h_2(W_{21}(A_1), W_{21}(B_1)) \\ &\quad \vee h_2(W_{22}(A_2), W_{22}(B_2))\} \quad \text{(by Lemma 3.1)} \\ &\le \max\{s_{11}h_1(A_1, B_1) \vee s_{12}h_2(A_2, B_2), \\ &\qquad s_{21}h_1(A_1, B_1) \vee s_{22}h_2(A_2, B_2)\} \\ &\le sh_1(A_1, B_1) \vee h_2(A_2, B_2) = s\tilde{h}((A_1, A_2), (B_1, B_2)) \\ &= s\tilde{h}(A, B). \end{aligned}$$

■

Corollary 3.5. *When $s<1$ there is a unique element*

$$A=(A_1, A_2, \ldots, A_N)\in \tilde{H}$$

such that

$$A_i = \bigcup_{j\in I(i)} W_{ij}(A_j) \qquad \text{for } i=1,2,\ldots,N,$$

i.e.,

$$W(A)=A.$$

We call A the attractor of the recurrent IFS.

Corollary 3.6 (Collage Theorem for recurrent IFS). *If $B\in\tilde{H}$ obeys*

$$\tilde{h}(B, W(B))\le \varepsilon>0,$$

then

$$\tilde{h}(B, A)\le \varepsilon/(1-s),$$

where A denotes the attractor of the recurrent IFS.

Remark. Although it is an elementary consequence of the contractivity of W this result has far-reaching consequences for image compression and analysis, just as in the case of the Collage Theorem for standard IFS, see [BS] for example.

To connect this with the original single space point map recurrent IFS, we have

Corollary 3.7. *Let $(K, w_i, p_{ij}, i, j=1,\ldots,N)$ be a recurrent IFS with K compact and the w_i's uniform contractions. Let A be the support of the unique stationary measure μ of Section 2. Then there exist unique compact sets $A_i\subset A$, $i=1,\ldots,N$ with $A=\bigcup_{i=1}^{N} A_i$ such that*

$$A_i = \bigcup_{j:p_{ji}>0} w_i(A_j), \qquad i=1,\ldots,N.$$

In terms of the random walk, the A_i's may be characterized as follows: for all x, $x\in A_i$ iff for every neighborhood G of x, for almost all trajectories x_0, $w_{i_1}x_0$, $w_{i_2}w_{i_1}x_0,\ldots$, we have $i_n=i$ and $w_{i_n}\cdots w_{i_1}x_0\in G$ for infinitely many n. In other words, to "see" A_i, just look at the points along a trajectory which end in map w_i.

Remark. Even if we are only interested in A itself, the invariance relation above for the decomposition is important; it is used, for example, to determine the fractal dimension of A in Section 4.

4. Fractal Dimension

4.1. Standard Recurrent IFS

Let $S \subset K$ be a subset of a compact metric space K with distance function d. Let $N(\varepsilon)$ denote the minimum number of balls of radius ε needed to cover S. Then the fractal dimension of S is defined to be

$$\dim(S) = \overline{\lim_{\varepsilon \to 0}} \frac{\ln N(\varepsilon)}{\ln 1/\varepsilon}.$$

Next let $S = (S_1, S_2, \ldots, S_N) \in H \times H \times \cdots \times H = H^N$, where, as in Section 3, H is the nonempty compact sets in K. Then we define

$$\dim(S) = \overline{\lim_{\varepsilon \to 0}} \left\{ \frac{\ln \sum_{k=1}^{N} N_k(\varepsilon)}{\ln 1/\varepsilon} \right\} = \max\{\dim S_k\} = \dim\left(\bigcup_{k=1}^{N} S_k \right),$$

where $N_k(\varepsilon)$ is the minimum number of balls needed to cover S_k.

Let $w_i: K \to K$ be Lipschitz maps which are uniform contractions, and (p_{ij}) an irreducible, row-stochastic matrix. In this section we are interested in the dimension of the attractor, so it is only the connection structure of the chain that concerns us. Let the connection matrix $C = (C_{ij})$ be defined by $C_{ij} = 1$ if $p_{ji} > 0$, 0 otherwise; that is, map w_i can follow map w_j iff $C_{ij} = 1$. We define the map $W: H^N \to H^N$ as in Section 3. This can be conveniently formulated in matrix notation as follows: let

$$W = (W_{ij}) = (C_{ij} W_i),$$

where W_i is the set map $W_i(S) = \{w_i(x): x \in S\}$ associated with w_i, and

$$C_{ij} W_i = \begin{cases} W_i & \text{if } C_{ij} = 1, \\ \varnothing & \text{if } C_{ij} = 0. \end{cases}$$

This matrix of set maps acts on H^N as we would expect by analogy with ordinary matrix multiplication: if $S = (S_1, \ldots, S_N)$ and $R = (R_1, \ldots, R_N) \in H^N$, then $WR = S$ means $S_i = \bigcup_{j=1}^{N} W_{ij}(R_j) = \bigcup_{j \in I(i)} W_i(R_j)$, where $I(i) = \{j: C_{ij} = 1\}$, as in Section 3.

The attractor is the unique element $A = (A_1, \ldots, A_N)$ of H^N satisfying $W(A) = A$ (Section 3). For example, for the connection diagram (Fig. 5) there corresponds

$$C = \begin{pmatrix} 1 & 1 \\ 1 & 0 \end{pmatrix}, \qquad W = \begin{pmatrix} W_1 & W_1 \\ W_2 & \varnothing \end{pmatrix},$$

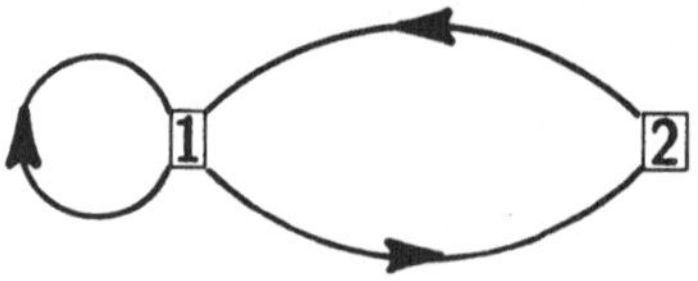

Fig. 5

and the attractor $A=(A_1, A_2)$ satisfies the invariance relations

$$A_1 = W_1(A_1) \cup W_1(A_2), \qquad A_2 = W_2(A_1).$$

We call the attractor *nonoverlapping* when $A_j \cap A_k = \varnothing$ for all j, $k \in I(i)$ such that $j \neq k$, $i = 1, 2, \ldots, N$. We call $w: K \to K$ a *similitude of contractivity* s if

$$d(wx, wy) = sd(x, y) \quad \text{for all } x, y.$$

Now we are ready for our first dimension result.

Theorem 4.1. *Let $w_i: K \to K$ be a similitude of contractivity s_i, $0 < s_i < 1$, for $i = 1, \ldots, N$. Let C be an $N \times N$ irreducible connection matrix, and for each $t \geq 0$, define a diagonal matrix $S(t) = \operatorname{diag}\{s_1^t, s_2^t, \ldots, s_N^t\}$. Let $A = (A_1, \ldots, A_N)$ be the attractor, whose existence follows by Corollary 3.5, and assume that A is nonoverlapping. Let d be the unique positive number such that 1 is an eigenvalue of $S(d)C$ of maximum modulus (see the Perron–Frobenius theorem below). Then* $\dim(A) = d$.

Proof. We need the

Perron–Frobenius Theorem [S]. *Let $M \geq 0$ be an irreducible square matrix. Then* (a) *$\rho(M)$, the spectral radius of M, is an eigenvalue of M and has strictly positive eigenvector y (i.e., $y_i > 0$ for all i), and* (b) *$\rho(M)$ increases if any element of M increases.*

Let $s_L = \min\{s_i\}$, $s_U = \max\{s_i\}$.

Since each A_j is compact, there is some $\varepsilon_0 > 0$ such that

$$(**) \qquad d(w_i(A_j), w_i(A_k)) > \varepsilon_0, \qquad \forall (j, k \in I(i) \text{ and } j \neq k), \quad i = 1, \ldots, N.$$

Let $N_i(\varepsilon)$ be the minimum number of ε-balls needed to cover A_i for $i = 1, \ldots, N$. From $A_i = \bigcup_{j \in I(i)} w_i(A_j)$ and $(**)$ and the fact that the maps are similitudes, we obtain the system of functional equations

$$(*) \qquad N_i(\varepsilon) = \bigcup_{j \in I(i)} N_j(\varepsilon / s_i) \qquad \text{for} \quad i = 1, \ldots, N,$$

for $0 < \varepsilon \leq \varepsilon_0 s_L$.

Let $x = (x_1, \ldots, x_N)$ be a strictly positive eigenvector of $S(d)C$ corresponding to the eigenvalue 1 (P–F theorem). We show that there are positive constants C_1 and C_2 such that

$$(***) \qquad C_1 \varepsilon^{-d} x_i \leq N_i(\varepsilon) \leq C_2 \varepsilon^{-d} x_i$$

for $i = 1, \ldots, N$, and $0 < \varepsilon \leq \varepsilon_0$. Pick $C_1 > 0$ and C_2 so that $(***)$ holds for $s_L \varepsilon_0 \leq \varepsilon \leq \varepsilon_0$. We proceed by induction. Assume $(***)$ holds for $s_U^n s_L \varepsilon_0 \leq \varepsilon \leq \varepsilon_0$. Then suppose $s_U^{n+1} s_L \varepsilon_0 \leq \varepsilon \leq s_L \varepsilon_0$; then $s_U^n s_L \varepsilon_0 \leq \varepsilon / s_i \leq \varepsilon_0$ and so

$$N_i(\varepsilon) = \sum_{j \in I(i)} N_j\left(\frac{\varepsilon}{s_i}\right) \leq C_2 \varepsilon^{-d} s_i^d \sum_{j \in I(i)} x_j = C_2 \varepsilon^{-d} x_i,$$

since x is an eigenvector, and in the same way, $C_1\varepsilon^{-d}x_i \le N_i(\varepsilon)$. Thus by induction we have it for $0<\varepsilon\le\varepsilon_0$. From this the theorem immediately follows. ■

4.2. *Recurrent Fractal Interpolation Functions*

Here we consider a generalization of the fractal interpolation functions introduced in [B]. Let $\{x_0, x_1, \ldots, x_N\}$ be a partition of $[0, 1]$ (i.e., $0=x_0<x_1<\cdots<x_N=1$) and $y_i\in\mathbf{R}$ for $i=0, 1, 2, \ldots, N$. We construct a "fractal function" which is continuous and whose graph contains the points (x_i, y_i), $i=0, 1, 2, \ldots, N$. For each $i=1, 2, \ldots, N$ let J_i denote the interval $[x_{i-1}, x_i]$ and choose J_i' to be an interval $[x_l, x_m]$ such that

$$(*) \qquad x_i - x_{i-1} < x_m - x_l.$$

Let $K=[0, 1]\times\mathbf{R}$ and define $w_i: K\to K$ to be an affine map of form

$$w_i: \begin{pmatrix} x \\ y \end{pmatrix} \to \begin{pmatrix} a_i & 0 \\ b_i & c_i \end{pmatrix}\begin{pmatrix} x \\ y \end{pmatrix} + \begin{pmatrix} d_i \\ e_i \end{pmatrix},$$

where we select a_i, b_i, c_i, d_i, e_i such that $|c_i|<1$ and such that w_i takes (x_l, y_l) to (x_{i-1}, y_{i-1}) and (x_m, y_m) to (x_i, y_i) (or (x_l, y_l) to (x_i, y_i) and (x_m, y_m) to (x_{i-1}, y_{i-1})). Note that $(*)$ implies $|a_i|<1$. Ket $a=\max\{|a_i|\}$, $b>\max\{|b_i|\}$, and $c=\max\{|c_i|\}$. We define a metric on K such that each w_i is contractive on K as follows: for $(x_1, y_1), (x_2, y_2)\in K$ define $d(x_1, y_1), (x_2, y_2)) = |x_1-x_2| + ((1-a)/2b)|y_1-y_2|$. It easily follows that each w_i is strictly contractive in this metric with contraction factor $s=\max\{(1+a)/2, c\}<1$. The recurrent structure is given by the connection matrix $C=(C_{ij})$ which is defined by $C_{ij}=1$ if $J_j\subset J_i'$ and $C_{ij}=0$ otherwise. Let $I(i)=\{j\colon C_{ij}=1\}$. We define the map $W\colon H^N\to H^N$ as in Subsection 4.1:

$$W=(W_{ij})=(C_{ij}W_i).$$

Corollary 3.5 implies that there is a unique $A=(A_1, A_2, \ldots, A_N)\in H^N$ such that $W(A)=A$. It follows as in [B] that $G=\bigcup_{i=1}^N A_i$ is the graph of a continuous function on $[0, 1]$. We call such a function a recurrent fractal interpolation function (r.f.i.f.). This construction is illustrated in Fig. 6. Here, for example, $J_1'=[x_0, x_2]=J_1\cup J_2$. The connection matrix C is given by

$$C=\begin{bmatrix} 1 & 1 & 0 & 0 & 0 & 0 \\ 0 & 0 & 1 & 1 & 0 & 0 \\ 1 & 1 & 0 & 0 & 0 & 0 \\ 0 & 0 & 0 & 0 & 1 & 1 \\ 0 & 0 & 1 & 1 & 0 & 0 \\ 0 & 0 & 0 & 0 & 1 & 1 \end{bmatrix}.$$

We now calculate the dimension of G.

Theorem 4.2. *Let f be an r.f.i.f. given by $\{w_i\colon i=1, \ldots, N\}$ with irreducible connection matrix C and graph G. Let $S(d)=\operatorname{diag}\{|c_1||a_1|^{d-1}, \ldots, |c_N||a_N|^{d-1}\}$ and let D be unique value so that $\rho(CS(D))=1$. If $\rho(CS(1))>1$ and there is some $k\in\{1, \ldots, N\}$ such that $\{(x_i, y_i)\colon x_i\in J_k'\}$ is not collinear then* $\dim(G)=D$, *otherwise* $\dim(G)=1$.

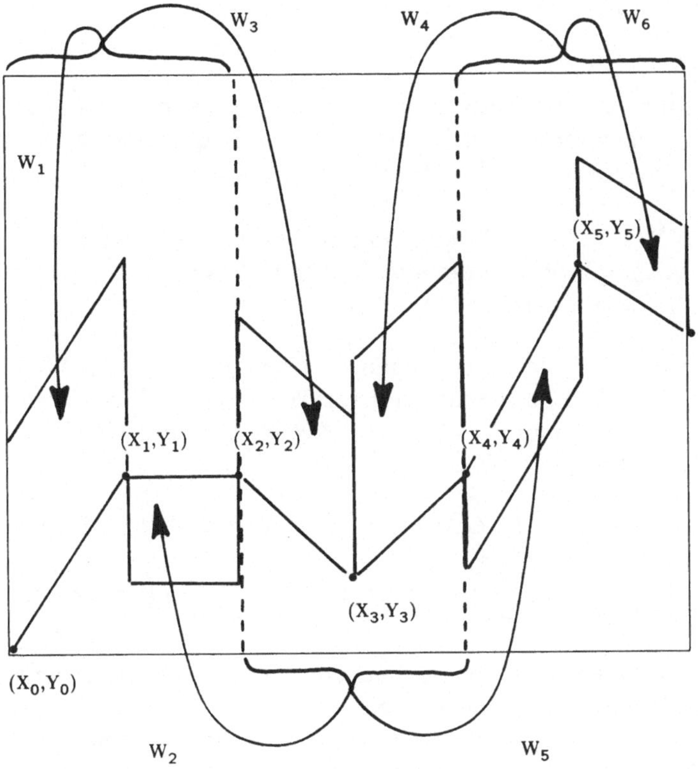

Fig. 6

Proof. As in [HM] and [BEHM] the main idea of the proof is to derive functional inequalities for $N(\varepsilon)$ which can then be used to estimate the behavior of $N(\varepsilon)$ as ε decreases to zero. The details of this proof follow closely those of Theorem 4 of [BEHM].

We first introduce a class of covers which allow us to relate covers of different sizes.

Definition. For $0<\varepsilon<1$, $\{\tau_l\}_{l=0}^{m}$ is called an *ε-partition* if

(a) $\tau_l \in (-\varepsilon/2, 1)$,
(b) $\varepsilon/2<\tau_{l+1}-\tau_l<\varepsilon$

for $l=1, 2, \ldots, m-1$. A cover $\mathscr{C}$ of G will be called an *ε-column cover of G* with associated ε-partition $\{\tau_l\}_{l=0}^{m}$ if there are positive integers $n_0, \ldots, n_m$ and real numbers $\xi_0, \ldots, \xi_m$ such that

$$\mathscr{C}=\{[\tau_l, \tau_l+\varepsilon]\times[\xi^l+(j_l-1)\varepsilon, \xi_l+j_l\varepsilon]: j_l=1, \ldots, n_k; l=0, 1, \ldots, m\}.$$

Note that $\mathscr{C}$ consists of $\sum_{l=0}^{m} n_l$ closed $\varepsilon\times\varepsilon$ squares arranged in $m+1$ columns. Let $|\mathscr{C}|$ denote the cardinality of $\mathscr{C}$ and define $\mathcal{N}^*(\varepsilon)=\min\{|\mathscr{C}|$: $\mathscr{C}$ is an ε-column cover of $G\}$ and let $\mathcal{N}(\varepsilon)$ be the minimum number of $\varepsilon\times\varepsilon$ squares $[a, a+\varepsilon]\times[b, b+\varepsilon]$, a, $b\in\mathbf{R}$ which cover G. Lemma 4.1 below shows that $\mathcal{N}^*(\varepsilon)$ can be used in the calculation of $\dim(G)$.

Lemma 4.1. $\mathcal{N}(\varepsilon)\leq\mathcal{N}^*(\varepsilon)\leq 2\mathcal{N}(\varepsilon)$, $\forall 0<\varepsilon<1$.

Proof. Clearly, $\mathcal{N}(\varepsilon)\leq\mathcal{N}^*(\varepsilon)$.

We introduce a third class of covers: a cover $\mathscr{C}$ of G will be called an *ε-nonoverlapping cover of G* if it consists of $\varepsilon\times\varepsilon$ squares with nonintersecting interiors of the form $[k\varepsilon,(k+1)\varepsilon]\times[y,y+\varepsilon]$ where $k\in\{0,1,\ldots,[\![1/\varepsilon]\!]\}$ and $y\in\mathbf{R}$.

Clearly, $\mathcal{N}^{**}(\varepsilon)\leq 2\mathcal{N}(\varepsilon)$. If $\mathscr{C}$ is a minimal ε-nonoverlapping cover of G then, since G is the graph of a continuous function, $\mathscr{C}$ is also an ε-column cover of G. Then $\mathcal{N}^*(\varepsilon)\leq\mathcal{N}^{**}(\varepsilon)\leq 2\mathcal{N}(\varepsilon)$. ■

Henceforth we only consider ε-column covers and we write $\mathcal{N}(\varepsilon)$ for $\mathcal{N}^*(\varepsilon)$. Let $A=(A_1,\ldots,A_N)$ be the attractor for $W=(C_{ij}W_i)$. Then $G=\bigcup_{i=1}^N A_i$ and A_i is the portion of G above $J_i=[x_{i-1},x_i]$. Let $N_i(\varepsilon)=\min\{|C|: C$ is an ε-column cover of $G\}$ for $i=1,\ldots,N$.

Lemma 4.2. *There exist P_i, $Q_i>0$, $i=1,\ldots,N$, such that for* $0<\varepsilon<1$.

$$(**)\qquad \left|\frac{c_i}{a_i}\right|\sum_{j\in I(i)}N_j\left(\frac{\varepsilon}{|a_i|}\right)-\frac{P_i}{\varepsilon}\leq N_i(\varepsilon)\leq\left|\frac{c_i}{a_i}\right|\sum_{j\in I(i)}N_j\left(\frac{\varepsilon}{|a_i|}\right)+\frac{Q_i}{\varepsilon}.$$

Proof. If $c_i=0$ then A_i is a line segment and we may choose P_i and Q_i proportional to the length of this line segment.

Now suppose $c_i\neq 0$, then w_i is invertible. Let $\mathscr{C}_i$ be a minimal ε-column cover of A_i and let R be a typical column R in $\mathscr{C}_i$ which consists of n $\varepsilon\times\varepsilon$ squares. Observe that $w_i^{-1}(R)$ is a parallelogram which can be covered by

$$\left[\!\!\left[n\left|\frac{a_i}{c_i}\right|+\left|\frac{b_i}{a_i}\right|+1\right]\!\!\right]$$

squares of sides $\varepsilon/|a_i|$ as shown in Fig. 7. Since there are at most $2|a_i|/\varepsilon+2$

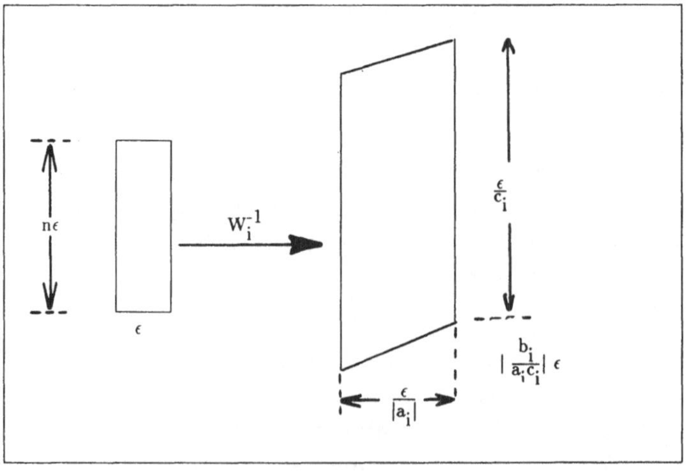

Fig. 7

columns in $\mathscr{C}_i$ we can find an $\varepsilon/|a_i|$-cover $\mathscr{D}$ of $\sum_{j\in I(i)} A_j$ such that

$$|\mathscr{D}| \le \left|\frac{a_i}{c_i}\right| N_i(\varepsilon) + 2\left(\frac{|a_i|}{\varepsilon}+1\right)\left(\left|\frac{b_i}{a_i}\right|+1\right).$$

Let $\mathscr{D}_j$ consist of the squares in $\mathscr{D}$ which meet $J_j\times\mathbf{R}=[x_{j-1}, x_j]\times\mathbf{R}$. Then $\mathscr{D}_j$ is an $\varepsilon/|a_i|$ cover of A_j and hence $|\mathscr{D}_j| \ge N_j(\varepsilon/|a_i|)$. Now if $q=|I(i)|$ and $M=\max_{x\in[0,1]}|f(x)|$ then we have $|\mathscr{D}|\ge\sum_{j\in I(i)}|\mathscr{D}_j| - 2q(2M/\varepsilon+1)$ (for we double count at most two columns at each of the $q-1$ points of x_j). Combining the above results yields

$$\begin{aligned} N_i(\varepsilon) &\ge \left|\frac{c_i}{a_i}\right|\left(|\mathscr{D}| - 2\left(\frac{|a_i|}{\varepsilon}+1\right)\left(\left|\frac{b_i}{a_i}\right|+1\right)\right) \\ &\ge \left|\frac{c_i}{a_i}\right| \sum_{j\in I(i)} N_j\left(\frac{\varepsilon}{|a_i|}\right) - \frac{P_i}{\varepsilon} \end{aligned}$$

for $0<\varepsilon<1$. Here

$$P_i = 2\left|\frac{c_i}{a_i}\right|\left[q(2m+1)+(|a_i|+1)\left(\left|\frac{b_i}{a_i}\right|+1\right)\right].$$

The upper bound follows in a similar way except that now we apply w_i to a minimal $\varepsilon/|a_i|$-column cover of A_j for $j\in I(i)$ to get an ε-column cover of A_i. ■

We proceed by induction on (**), however, to get the induction started we need the following lemma.

Lemma 4.3. *If $\lambda=\rho(CS(1))>1$ and there is some $k\in\{1,\dots,N\}$ such that $\{(x_j, y_j): x_j\in J'_k\}$ is not collinear, then $\lim_{\varepsilon\to 0}\varepsilon N_i(\varepsilon)=\infty$ for $i=1,2,\dots,N$.*

Proof. From the noncollinearity of the interpolation points above J'_k and the irreducibility of the connection matrix C it follows that A_i is not a line segment for $i\in\{1,\dots,N\}$. Thus we can find points $P_i(\alpha_i,\beta_i)$, $Q_i(\alpha'_i,\beta'_i)$, $R(\alpha''_i,\beta''_i)$, $\alpha_i<\alpha'_i<\alpha''_i$, on the graph of f which are not collinear and which are in the interior of $J_i\times\mathbf{R}$. Let

$$s_i = \left|\beta'_i - \left(\beta_i + \left(\frac{\beta''_i-\beta_i}{\alpha''_i-\alpha_i}\right)(\alpha'_i-\alpha_i)\right)\right|$$

as shown in Fig. 8. Let $\delta=\min\{|\alpha''_i-\alpha_{i+1}|: i=1,\dots,N-1\}>0$. Let $v=(v_1,\dots,v_N)$ be a strictly positive eigenvector of $CS(1)$ with eigenvalue λ such that $v_i\le s_i$ for $i=1,\dots,N$. Since f is continuous we must have $N_i(\varepsilon)\ge s_i/\varepsilon\ge v_i/\varepsilon$. Let $\mathbf{a}=\min\{|a_k|: k=1,\dots,N\}$. It follows from $A_i=\bigcup_{j\in I(i)} w_i(A_j)$ that if $\varepsilon<\delta\mathbf{a}$ then

$$N_i(\varepsilon)\ge|c_i|\left(\sum_{j\in I(i)} s_j\right)\Big/\varepsilon \ge \frac{|c_i|}{\varepsilon}\sum_{j\in I(i)} v_j = \frac{1}{\varepsilon}(CS(1)v)_i = \frac{\lambda}{\varepsilon}v_i.$$

An induction gives $N_i(\varepsilon)\ge\lambda^n(v_i/\varepsilon)$ which proves the lemma. ■

Now back to (**). If $\lambda=\rho(S(1)C)>1$ then there is a unique $D>1$ such that $\rho(S(D)C)=1$. Let v be a strictly positive eigenvector of $S(1)C$ with eigenvalue

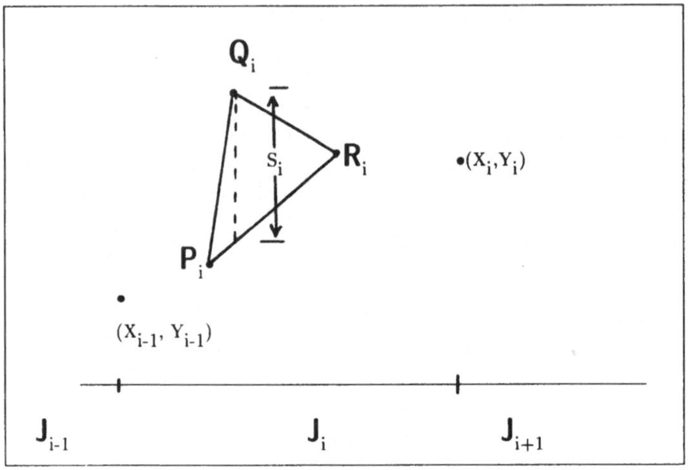

Fig. 8

λ and let w be a strictly positive eigenvector of $S(D)C$ with eigenvalue 1. Choose $P>0$ and $Q>0$ so that $Pv_i \geq P_i$ and $Qv_i \geq Q_i$ for $i=1,\ldots,N$. Then (**) becomes

$$(***)\qquad \left|\frac{c_i}{a_i}\right| \sum_{j\in I(i)} N_j\left(\frac{\varepsilon}{|a_i|}\right) - P\frac{v_i}{\varepsilon} \leq N_i(\varepsilon) \leq \left|\frac{c_i}{a_i}\right| \sum_{j\in I(i)} N_j\left(\frac{\varepsilon}{|a_i|}\right) - Q\frac{v_i}{\varepsilon}.$$

We first consider the system of functional equalities associated with (***):

$$g_i(\varepsilon) = \left|\frac{c_i}{a_i}\right| \sum_{j\in I(i)} g_j\left(\frac{\varepsilon}{|a_i|}\right) + R\frac{v_i}{\varepsilon} \qquad \text{for} \quad i=1,\ldots,N$$

which, we can verify, has the system of solutions

$$\varphi_i(R,\gamma,\varepsilon) = \gamma\varepsilon^{-D}w_i + \frac{R}{1-\lambda}\varepsilon^{-1}v_i \qquad \text{for} \quad i=1,\ldots,N,$$

where γ is an arbitrary constant and R will either equal Q or $-P$. Pick γ_1 large enough so that

$$N_i(\varepsilon) \leq \varphi_i(Q,\gamma_1,\varepsilon)$$

for $i=1,\ldots,N$ and $\mathbf{a} \leq \varepsilon \leq 1$ where, as before, $\mathbf{a} = \min\{|a_i| : i=1,\ldots,N\}$. If $a\mathbf{a} \leq \varepsilon \leq \mathbf{a}$ then $\mathbf{a} \leq \varepsilon/|a_i| \leq 1$ so by (***)

$$\begin{aligned} N_i(\varepsilon) &\leq \left|\frac{c_i}{a_i}\right| \sum_{j\in I(i)} N_j\left(\frac{\varepsilon}{|a_i|}\right) - Q\frac{v_i}{\varepsilon} \\ &\leq \left|\frac{c_i}{a_i}\right| \sum_{j\in I(i)} \varphi_j\left(Q,\gamma_1,\frac{\varepsilon}{|a_i|}\right) - Q\frac{v_i}{\varepsilon} \leq \varphi_i(Q,\gamma_1,\varepsilon) \end{aligned}$$

for $i=1,\ldots,N$ and $a\mathbf{a} \leq \varepsilon \leq 1$. It follows by induction that

$$N_i(\varepsilon) \leq \varphi_i(Q,\gamma_1,\varepsilon)$$

for $a^n\mathbf{a} \le \varepsilon \le 1$, $n = 1, 2, \ldots$, and since $a < 1$ this must hold for $0 < \varepsilon \le 1$. Thus

$$\dim(G) \le \lim_{\varepsilon \to 0} \frac{\log\left(\sum_{i=1}^{N} \varphi_i(Q, \gamma_1, \varepsilon)\right)}{\log(1/\varepsilon)} = \max\{D, 1\}.$$

Assume now that the hypotheses of Lemma 4.3 are satisfied. Let ε_0 be small enough so that $\varepsilon N_i(\varepsilon) > (R/(1-\lambda))\varepsilon^{-1} v_i$ for $\varepsilon < \varepsilon_0$ and all $i = 1, \ldots, N$. Then we can pick $\gamma_2 > 0$ such that

$$N_i(\varepsilon) \ge \varphi_i(Q, \gamma_2, \varepsilon)$$

for $\mathbf{a}\varepsilon_0 \le \varepsilon \le \varepsilon_0$. As for the upper bound, we use (***) to show that

$$N_i(\varepsilon) \ge \varphi_i(Q, \gamma_2, \varepsilon)$$

for $0 < \varepsilon < \varepsilon_0$ and hence

$$\dim G = \max\{D, 1\} = D.$$

If $\{(x_j, y_j)\colon x_j \in J_i'\}$ is collinear for all $i = 1, \ldots, N$, then G consists of a finite number of line segments so that $\dim G = 1$.

Since G is the graph of function we must have $\dim G \ge 1$. In general we have $\dim G \le \max\{D, 1\}$. If $\lambda \le 1$ then $D \le 1$ so that $\dim G = 1$. ■

5. Examples: Julia Sets, Boundaries of IFS Attractors, Fractal Interpolation Functions, etc.

5.1. A Julia Set Example

Let $T\colon \hat{\mathbf{C}} \to \hat{\mathbf{C}}$ be defined by $Tz = (z-\lambda)^2$ where $\lambda \in \mathbf{R}$ is a parameter such that $0.75 < \lambda < 1.25$, and where $\hat{\mathbf{C}} = \mathbf{C} \cup \infty$. Let w_+ and w_- denote two branches of the inverse of T, defined so that w_+ lies in the upper half-plane union the positive real axis and w_- lies in the lower half-plane union the negative real axis; that is, if $z = \lambda + \Gamma\, e^{i\theta}$ with $0 \le \theta \le 2\pi$ and $\Gamma > 0$, then

$$w_+ z = \sqrt{\Gamma}\, e^{i\theta/2} \quad \text{and} \quad w_- z = \sqrt{\Gamma}\, e^{i\theta/2 + i\pi}.$$

Let $\mathcal{O}$ denote any sufficiently small neighborhood of the pair of points $\{\lambda + \frac{1}{2} \pm \sqrt{\lambda + \frac{1}{4}}\}$, which is the attractive two-cycle for the map T for $\frac{3}{4} < \lambda < \frac{5}{4}$, see [BGH], for example. Let $K = \hat{\mathbf{C}} \setminus \mathcal{O}$. Then $\{k, w_+, w_-\}$ is an IFS. Although as it stands, with respect to the Euclidean metric in $\hat{\mathbf{C}}$, this IFS is not hyperbolic, it does possess a unique attractor; and, because the critical point of T does not lie on the attractor of the IFS, it is possible to treat the system essentially as though it were strictly contractive: that is, there exists a metric such that the Collage Theorem applies.

The attractor for the IFS is the Julia set J for T. It looks something like Fig. 9. J is characterized by $J = w_+(J) \cup w_-(J)$, as promised by Corollary 2. Note that w_+ applied to J yields the part of J which lies in the upper half-plane. w_+ unwraps the set about zero, cut along the branch cut, so that it lies in the upper half-plane, then shifts it to the right by one bubble. The attractive two-cycle resides in the interior of the two shaded bubbles.

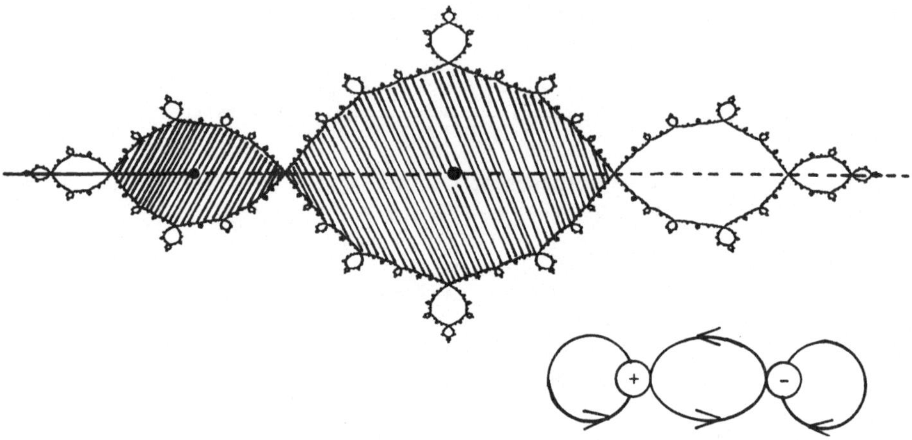

Fig. 9

Suppose we are interested in computing and studying not the whole of J, but only the boundary of the two-cycle. Then we need to consider the recurrent IFS whose diagram is shown in Fig. 10. This follows from Theorem 3.4 and its corollaries once we notice how the boundary of the two-cycle is mapped into itself (Fig. 11).

Another example of this type concerns the attractive three-cycle which exists when $\lambda = 1.75$ or thereabouts. When T admits an attractive real three-cycle, there exists a part of the Julia set $J_R = J \cap \mathbf{R}$, which contains cycles of all orders and is what is referred to by Li and Yorke [LiY] in their celebrated paper "Period Three Implies Chaos." It is possible to consider the closure of the analytic continuation of this set of cycles to all values of λ: J_R is the attractor for the recurrent IFS (Fig. 12), see [BGH], for example. Many examples involving interesting pieces of Julia sets may be constructed.

5.2. Boundaries of Attractors of IFS

A current problem in standard IFS theory is how to calculate the fractal dimensions of boundaries of IFS attractors. Here we illustrate by means of an example that the recurrent theory provides a key to this problem.

Let $K \subset \mathbf{R}^2$ be a large bounded disk, and consider the IFS $\{K, w_1, w_2, w_3\}$

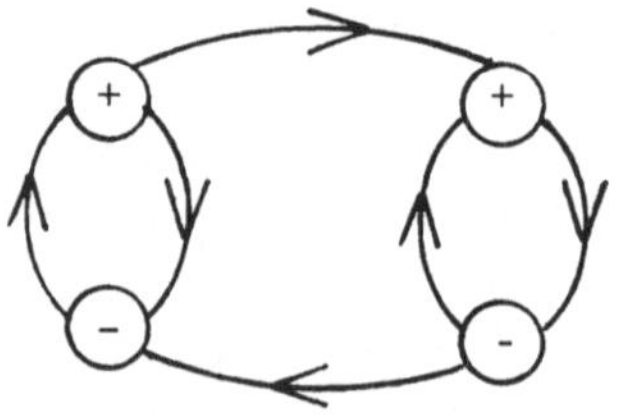

Fig. 10

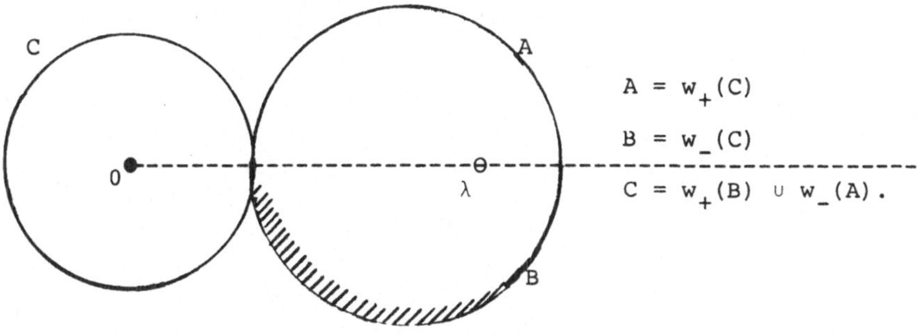

Fig. 11

where each w_i is a similitude,

$$w_i\begin{pmatrix}x\\y\end{pmatrix}=s_i\begin{pmatrix}\cos\theta_i & -\sin\theta_i\\ \sin\theta_i & \cos\theta_i\end{pmatrix}\begin{pmatrix}x\\y\end{pmatrix}+\begin{pmatrix}a_i\\b_i\end{pmatrix},$$

$0<s_i<1$, $0\le\theta_i<2\pi$, a_i, $b_i\in\mathbf{R}$, $i=1, 2, 3$, chosen as illustrated in Fig. 13.

$$\begin{array}{lll} w_1(A)=A, & w_1(C)=B, & w_1(E)=F,\\ w_2(A)=C, & w_2(C)=D, & w_2(E)=B,\\ w_3(A)=E, & w_3(C)=F, & w_3(E)=D. \end{array}$$

It is shown in [BEHM] that attractors of such IFS are connected. Let S denote the attractor. It is "just touching" and self-similar.

Remark. The fractal dimension D of S is the same as its Hausdorff-Besicovitch dimension and is given by solving $s_1^D+S_2^D+s_3^D=1$. If $s_1=s_2=s_3=\frac{1}{2}$, then $D=(\log 3)/(\log 2)$ and S is a Sierpinski gasket.

We are concerned with the characterization and fractal dimension of the outer boundary ∂S of S, namely the boundary of the component of the complement of S which contains 0. When $s_1=s_2=s_3=\frac{1}{2}$ and S is a Sierpinski gasket, ∂S is just a triangle. In general it is more complicated.

The point to realize about outer boundaries of IFS attractors is $w_i^{-1}(\partial S)\subset S$ for each i; that is, all points on the outer boundary "come from" the outer boundary under the IFS maps. This suggests that we may sometimes be able to find a combinatorial IFS description of the boundary, as we can in the present case.

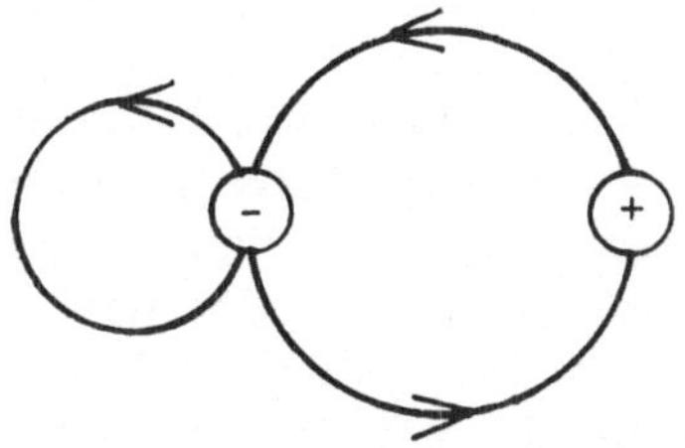

Fig. 12

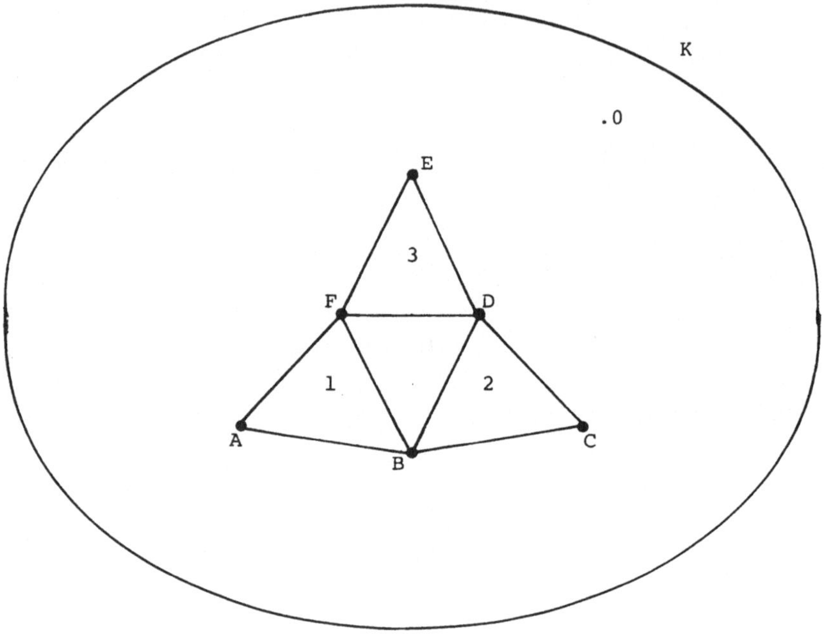

Fig. 13

Let AB denote the part of ∂S connecting A to B (via points whose codes commence with 1), BC denote the part of ∂S connecting B to C (via points whose codes commence with 2), and so on. Let AC denote $AB \cup BC$, $CE = CD \cup DE$, and $EA = EF \cup FA$. Then we observe that

$$EQ = w_1(EA) \cup w_3(AC),$$

$$AC = w_1(AC) \cup w_2(EA),$$

$$CE = w_2(AC) \cup w_3(EA).$$

Let $H^3 = H \times H \times H$, where H is the collection of nonempty compact sets in K. Then the recurrent IFS corresponding to ∂S can be represented by $W: H^3 \to H^3$ given by

$$W = \begin{pmatrix} w_1 & w_3 & \varnothing \\ w_2 & w_1 & \varnothing \\ w_3 & w_2 & \varnothing \end{pmatrix}.$$

∂S is then the projection of the attractor (which lies in three planes) onto one plane, as symbolized in Fig. 14. This produces the whole of ∂S. Notice that the IFS is not recurrent, strictly speaking, when viewed as a process in one copy of K: the process would then leak away via EC. However, the part of ∂S given by $\partial S_1 = AE \cup AC$ *is* recurrent in the sense of Section 2; we can get from any section of ∂S_1 to any other by following the maps w_1, w_2, and w_3 in the right order. ∂S_1 is the attractor for the recurrent IFS $\{K, w_1, w_2, w_3\}$ corresponding to a directed

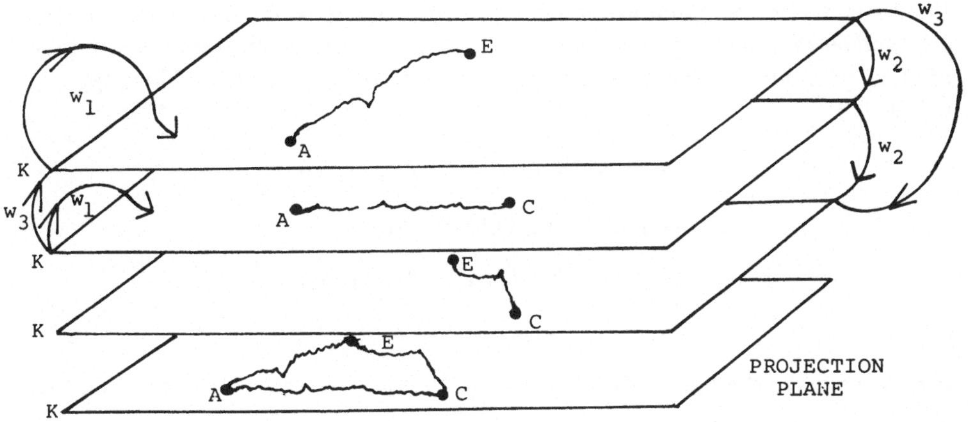

Fig. 14

graph of the form shown in Fig. 15. If $w_4 = w_1$, and if p_{ij} is the conditional probability of applying map j given that map i was applied at the previous step, then the transition matrix has the structure

$$\begin{bmatrix} p_{11} & p_{12} & 0 & 0 \\ 0 & 0 & p_{23} & p_{24} \\ p_{31} & p_{32} & 0 & 0 \\ 0 & 0 & p_{43} & p_{44} \end{bmatrix},$$

where the p_{jk}'s are nonzero.

Clearly, the fractal dimension of ∂S is the same as that of ∂S_1, because ∂S is the union of ∂S_1 with two affine images of sections of ∂S_1, such that the intersections of these pieces are of fractal dimension zero. The fractal dimension of ∂S_1 can be computed using the theory of Section 4. We give a sketch of the main argument here. Note that ∂S_1 is made of four pieces:

$\partial S_1 =$ 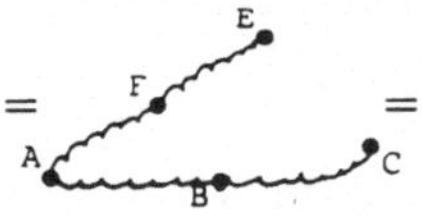 $= AB \cup BC \cup EF \cup FA.$

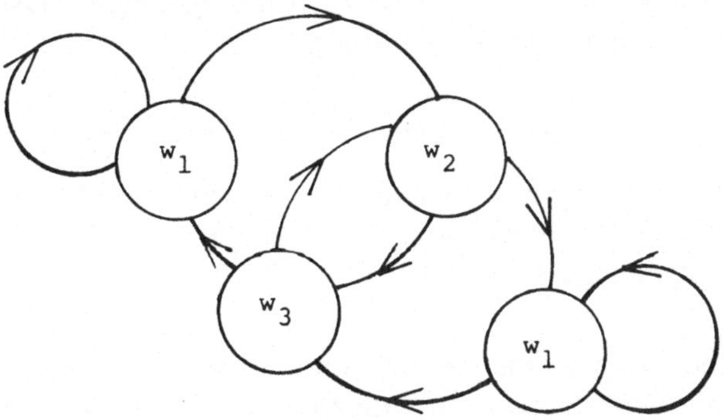

Fig. 15

Let $N(\varepsilon)$ denote the minimum number of disks of radius $\varepsilon>0$ required to cover ∂S_1; and let $N_{AB}(\varepsilon)$ (resp. $N_{BC}(\varepsilon)$, $N_{FE}(\varepsilon)$, $N_{FA}(\varepsilon)$) denote the minimum number of disks of radius $\varepsilon>0$ required to cover AB (resp. BC, FE, FA). Then approximately

$$N_{FA}(\varepsilon)\simeq N_{FA}(\varepsilon/s_1)+N_{FE}(\varepsilon/s_1),$$

$$N_{FE}(\varepsilon)\simeq N_{AB}(\varepsilon/s_3)+N_{BC}(\varepsilon/s_3),$$

$$N_{BC}(\varepsilon)\simeq N_{FA}(\varepsilon/s_2)+N_{FE}(\varepsilon/s_2),$$

$$N_{AB}(\varepsilon)\simeq N_{AB}(\varepsilon/s_1)+N_{BC}(\varepsilon/s_1).$$

If now we make the ansatz $N_{AB}(\varepsilon)\sim a_{AB}\varepsilon^{-D}$, $N_{BC}(\varepsilon)\sim a_{BC}\varepsilon^{-D}$, and so on, then we find

$$a_{FA}=(a_{FA}+a_{FE})s_1^D,$$

$$a_{FE}=(a_{FA}+a_{BC})s_3^D,$$

$$a_{BC}=(a_{FA}+a_{FE})s_2^D,$$

$$a_{AB}=(a_{AB}+a_{BC})s_1^D.$$

The fractal dimension D is given by finding the unique solution $(a_{FA}, a_{FE}, a_{AB}, a_{BC}, D)$ of this set of equations with all components positive. To say this another way, D is the unique positive number such that the matrix

$$\begin{bmatrix} s_1^D & 0 & 0 & 0 \\ 0 & s_2^D & 0 & 0 \\ 0 & 0 & s_3^D & 0 \\ 0 & 0 & 0 & s_1^D \end{bmatrix}\begin{bmatrix} 1 & 1 & 0 & 0 \\ 0 & 0 & 1 & 1 \\ 1 & 1 & 0 & 0 \\ 0 & 0 & 1 & 1 \end{bmatrix},$$

has eigenvalue one corresponding to positive eigenvector

$$(a_{FA}, a_{FE}, a_{BC}, a_{AB})^+.$$

Notice that if $s_1=s_2=s_3=\frac{1}{2}$ then the solution is $a_{FA}=a_{FE}=a_{BC}=a_{AB}=1$ and $D=1$. We independently confirm this result by noting that the classical Sierpinski gasket has outer boundary equal to a triangle.

5.3. Two Computer-Graphical Examples

Example 1. The four images shown in Fig. 16 correspond to a single recurrent IFS on $H^4=H\times H\times H\times H$ where K is a rectangle in the Euclidean plane, and H is the nonempty compact subsets of K.

Example 2. The "opposed-alternate" and "alternate-opposed" fern images shown in Fig. 17 were constructed with the aid of the recurrent IFS Collage Theorem.

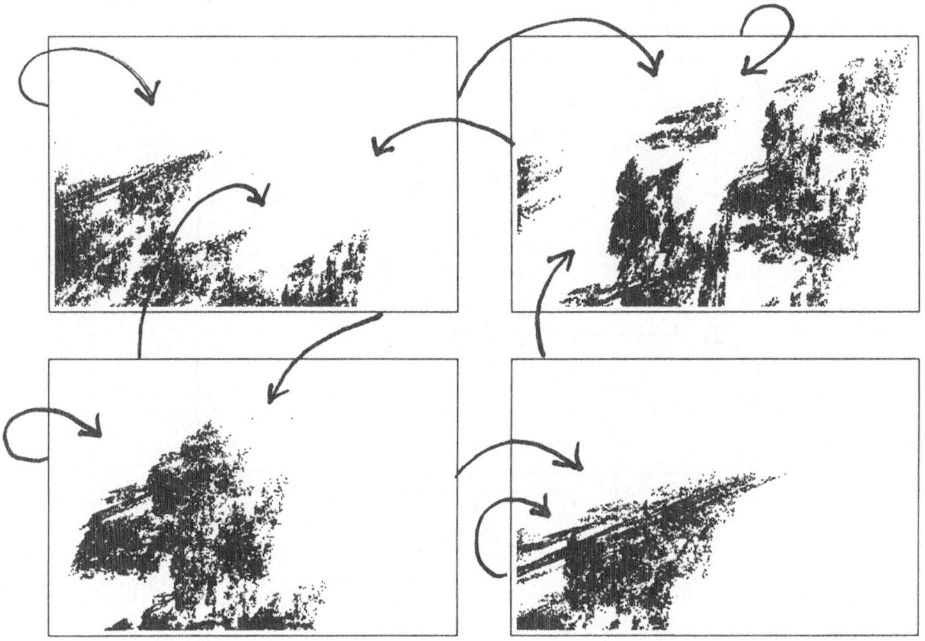

Fig. 16

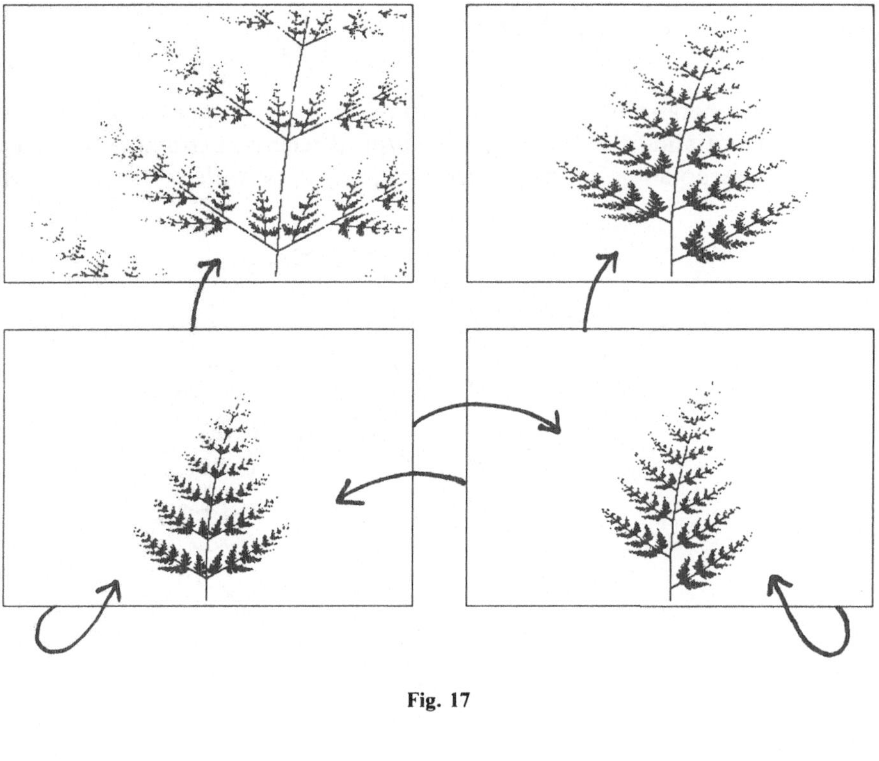

Fig. 17

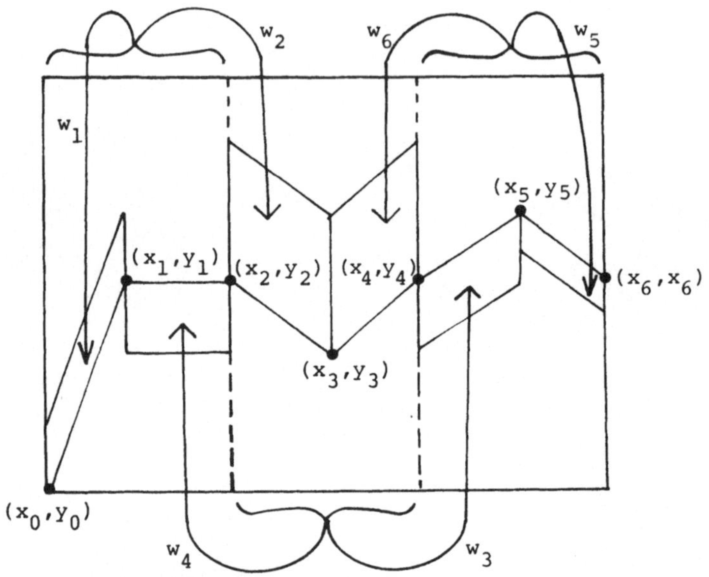

Fig. 18

5.4. *Recurrent Fractal Interpolation Functions*

We work, for example, in $\mathbf{R}^2$. Each map is of the special form $w_i:\mathbf{R}^2\to\mathbf{R}^2$,

$$w_i\begin{pmatrix}x\\y\end{pmatrix}=\begin{pmatrix}a_i & 0\\ b_i & c_i\end{pmatrix}\begin{pmatrix}x\\y\end{pmatrix}+\begin{pmatrix}e_i\\f_i\end{pmatrix}$$

and nonlinear generalizations of this structure. a_i, b_i, c_i, e_i, f_i are real constants. We choose $|a_i|<1$ and $|c_i|<1$, as in [B]. A typical recurrent structure associated with a set of interpolation points (x_i, y_i), $i=0, 1, \ldots, N$, with $x_0<x_1<\cdots<x_N$,

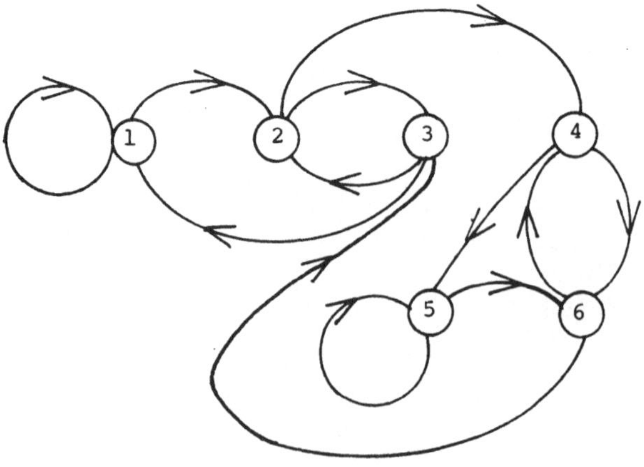

Fig. 19

is symbolized in Fig. 18.

$$
\begin{aligned}
&w_1(x_0, y_0) = (x_0, y_0), && w_3(x_2, y_2) = (x_4, y_4), && w_5(x_6, y_6) = (x_6, y_6),\\
&w_1(x_2, y_2) = (x_1, y_1), && w_3(x_4, y_4) = (x_5, y_5), && w_5(x_4, y_4) = (x_4, y_4),\\
&w_2(x_0, y_0) = (x_2, y_2), && w_4(x_2, y_2) = (x_1, y_1), && w_6(x_4, y_4) = (x_4, y_4)\\
&w_2(x_2, y_2) = (x_3, y_3), && w_4(x_4, y_4) = (x_2, y_2), && w_6(x_6, y_6) = (x_3, y_3).
\end{aligned}
$$

The corresponding directed graph is shown in Fig. 19. The attractor of such a recurrent IFS is the graph of a function which passes through the interpolation points.

References

[B] M. F. Barnsley (1986): *Fractal functions and interpolation.* Constr. Approx., **2**:303-329.

[BD] M. F. Barnsley, S. Demko (1985): *Iterated function systems and the global construction of fractals.* Proc. Roy. Soc. London Ser. A, **399**:243-275.

[BE] M. F. Barnsley, J. Elton (1988): *A new class of Markov processes for image encoding.* Adv. in Appl. Probab., **20**:14-32.

[BEHL] M. F. Barnsley, V. Ervin, D. P. Hardin, J. Lancaster (1986): *Solution of an inverse problem for fractals and other sets.* Proc. Nat. Acad. Sci. U.S.A., **83**:1975-1977.

[BEHM] M. F. Barnsley, J. Elton, D. Hardin, P. Massopust (preprint): Hidden Variable Fractal Interpolation Functions.

[BGH] M. F. Barnsley, J. S. Geronimo, A. N. Harrington (1984): *Geometry and combinatorics of Julia sets of real quadratic maps.* J. Statist. Phys., **37**:51-92.

[BS] M. F. Barnsley, A. D. Sloan (1986) (preprint): *Image compression.*

[Bed] T. Bedford *Dimension and dynamics for fractal recurrent sets.*

[BA] M. Berger, Y. Amit (preprint): *Products of random affine maps.*

[D] F. M. Dekking (1982): Recurrent Sets: A Fractal Formalism. Delft University of Technology.

[Dj] J. Dugundji (1966): Topology. Boston: Allyn and Bacon, p. 253.

[E] J. Elton (1987): *An ergodic theorem for iterated maps.* Ergodic Theory Dynamical Systems, **7**:481-488.

[F] W. Feller (1957): An Introduction to Probability Theory and Its Applications. London: Wiley.

[H] J. Hutchinson (1981): *Fractals and self-similarity.* Indiana Univ. Math. J., **30**:731-747.

[HM] D. P. Hardin, P. Massopust (1986): *Dynamical systems arising from iterated function systems.* Comm. Math. Phys., **105**:455-460.

[Kr] U. Krengel (1985): Ergodic Theorems, New York: de Gruyter.

[LiY] T. Li, J. A. Yorke (1975): *Period three implies chaos.* Amer. Math. Monthly, **82**:985-992.

[S] E. Seneta (1973): Non-negative Matrices. New York: Wiley.

M. F. Barnsley
School of Mathematics
Georgia Institute of Technology
Atlanta
Georgia 30332
U.S.A.

J. H. Elton
School of Mathematics
Georgia Institute of Technology
Atlanta
Georgia 30332
U.S.A.

D. P. Hardin
School of Mathematics
Georgia Institute of Technology
Atlanta
Georgia 30332
U.S.A.

Constr. Approx. (1989) 5: 33–48

CONSTRUCTIVE APPROXIMATION

Hölder Exponents and Box Dimension for Self-Affine Fractal Functions

Tim Bedford

Abstract. We consider some self-affine fractal functions previously studied by Barnsley *et al.* The graphs of these functions are invariant under certain affine scalings, and we extend their definition to allow the use of nonlinear scalings. The Hölder exponent, h, for these fractal functions is calculated and we show that there is a larger Hölder exponent, h_λ, defined at almost every point (with respect to Lebesgue measure). For a class of such functions defined using linear affinities these exponents are related to the box dimension D_B of the graph by $h \le 2 - D_B \le h_\lambda$.

1. Introduction

Barnsley [Ba], Barnsley and Harrington [BH], and Hardin and Massopust [HM] have introduced the idea of interpolation using fractal functions. It is easy to find a fractal function passing through a given finite set of points and even if we restrict ourselves to fractal functions defined using linear contractions (we give the full definition below) there are a number of free parameters left in the construction. It is known that these parameters can be chosen to make the box dimension (otherwise known as the capacity or logarithmic density—see [Tr]) of the fractal curve equal to any given number (strictly) between 1 and 2. The aim here is to show that we can alternatively choose the free parameters to make the Hölder exponent or the almost everywhere Hölder exponent equal to any given number in (0, 1).

In Section 2 we give some definitions and basic results. Section 3 deals with the calculations of exponents for a class of self-affine functions defined using linear scalings. In Section 4 a more general definition of self-affine functions is given which allows the use of nonlinear contractions. The extra theory needed to deal with the calculations in this case is given, but the main ideas of the proof are identical to those used in Section 3.

Date received: December 15, 1986. Communicated by Michael F. Barnsley.
AMS classification: 26A30, 41A30, 58F11.
Key words and phrases: Fractals, Self-affine, Hölder exponents, Box dimension, Gibbs measures.

2. Preliminary Definitions

Here we give the construction of self-affine functions employed by Barnsley and co-workers in [Ba], [BH], and [HM]. In addition we also use the well-known coding of points on the fractal graph by infinite sequences of symbols. Given a collection of points (x_0, y_0), $(x_2, y_2), \ldots, (x_n, y_n)$, where without loss of generality we take $x_0 = 0$ and $x_n = 1$, we construct a self-affine curve through $(x_0, y_0), \ldots, (x_n, y_n)$ as follows. Choose for each $i = 0, \ldots, n-1$ a linear mapping $\varphi_i : \mathbf{R}^2 \to \mathbf{R}^2$ of the plane of the form

(1) $$\begin{pmatrix} x \\ y \end{pmatrix} \to \begin{pmatrix} a_i & 0 \\ b_i & c_i \end{pmatrix} \begin{pmatrix} x \\ y \end{pmatrix} + \begin{pmatrix} v_i^0 \\ v_i^1 \end{pmatrix}$$

with

(2) $$0 < a_i < c_i < 1$$

and the further properties that

(3) $$\varphi_i((x_0, y_0)) = (x_i, y_i)$$

and

(4) $$\varphi_i((x_n, y_n)) = (x_{i+1}, y_{i+1}).$$

Suppose that each φ_i is a contraction mapping. Then by a theorem of Hutchinson [Hu] there is a unique compact set E in the plane with the property that

(5) $$E = \bigcup \varphi_i(E).$$

If the φ_i are not contractions in the Euclidean metric on $[0, 1] \times \mathbf{R}$ then we can find (see [HM]) an equivalent metric in which they are. Alternatively we can replace the family $\{\varphi_i\}$ by the family

$$\{\varphi_{x(1)} \circ \cdots \circ \varphi_{x(m)} | 0 \le x(1), \ldots, x(m) < n\}$$

which for large enough m are contractions on $[0, 1] \times \mathbf{R}$. Then there exists a compact nonempty set E with

$$\bigcup \varphi_{x(1)} \circ \cdots \circ \varphi_{x(m)}(E) = E$$

and so also $E = \bigcup \varphi_i(E)$. Furthermore, by conditions (3) and (4) E contains the points $(x_0, y_0), \ldots, (x_n, y_n)$. In fact we will see that E is the graph of a function $f : I \to \mathbf{R}$. For convenience we write $[0, 1] = I$. As each φ_i preserves the set of vertical lines $\{(x, y) : x = c\}$ there is an induced map ψ_i of I given by $\psi_i(x) = a_i x + v_i^0$. Notice that $\{\psi_i\}$ is a collection of contraction maps with

$$I = \bigcup \psi_i(I).$$

Let us establish some notation and basic definitions. The *full shift on n symbols* is the space

$$\Sigma_n = \{0, \ldots, n-1\}^{\mathbf{N}} = \{\mathbf{x} = x(1)x(2) \cdots : x(j) \in \{0, \ldots, n-1\}\},$$

together with the shift map $\sigma : \Sigma_n \to \Sigma_n$ defined by

$$\sigma(x(1)x(2)x(3) \cdots) = x(2)x(3) \cdots.$$

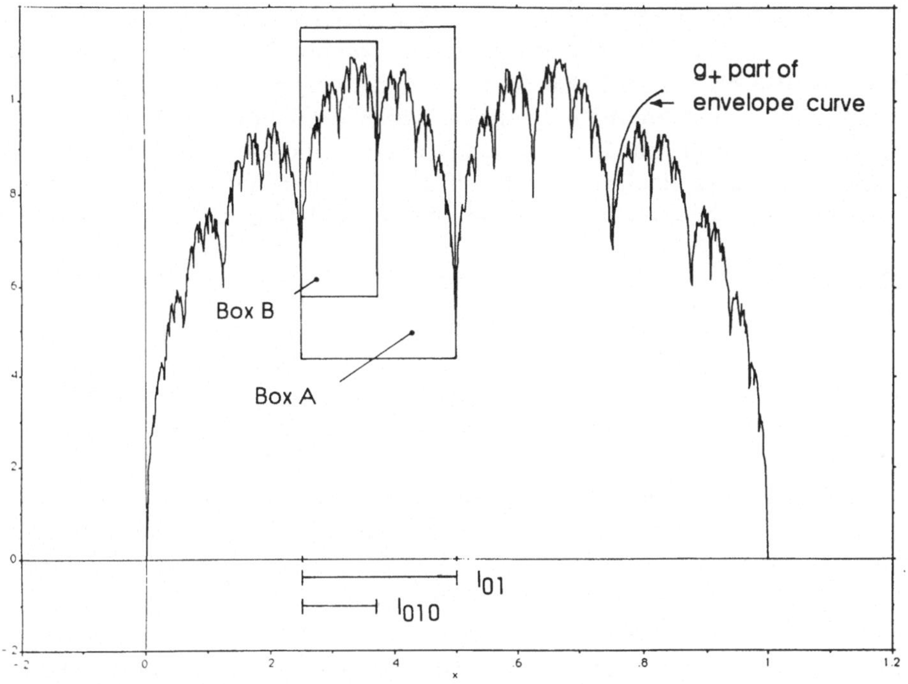

Fig. 1. The portion of the curve contained in box A is E_{01}. Box B contains E_{010}.

A finite word $x(1)\cdots x(m)$ is denoted $\mathbf{x}(m)$. The composition $\varphi_{x(1)}\cdots\varphi_{x(m)}$ is written $\varphi_{\mathbf{x}(m)}$ and similarly we write $\psi_{\mathbf{x}(m)}=\psi_{x(1)}\cdots\psi_{x(m)}$, $I_{\mathbf{x}(m)}=\psi_{\mathbf{x}(m)}(I)$, and $E_{\mathbf{x}(m)}=\varphi_{\mathbf{x}(m)}(E)$. The set $I_{\mathbf{x}(m)}$ is called an *m-interval*, and the sets $E_{\mathbf{x}(m)}$ and

$$C_{\mathbf{x}(m)}=\{\mathbf{z}\in\Sigma_n: \mathbf{z}(m)=\mathbf{x}(m)\}$$

are called *m-cylinders*. Figure 1 gives some examples of $E_{\mathbf{x}(m)}$ and $I_{\mathbf{x}(m)}$ for a particular curve.

Given any infinite word $\mathbf{x}=x(1)\cdots x(m)\cdots$, it is clear that diam $I_{\mathbf{x}(m)}$ and diam $E_{\mathbf{x}(m)}$ converge to zero as m tends to infinity. Hence corresponding to each infinite word $\mathbf{x}=x(1)\cdots x(m)\cdots$ there are points $x\in\mathbf{R}$ and $(x,y)\in\mathbf{R}^2$ such that

$$\{x\}=\bigcap_m I_{\mathbf{x}(m)} \quad\text{and}\quad \{(x,y)\}=\bigcap_m E_{\mathbf{x}(m)}.$$

We define $\pi:\Sigma_n\to I$ by $\pi(\mathbf{x})=x$ (and use the notational convention that $\pi(\mathbf{x})=x$). Notice that $\psi_j(\pi(x(1)x(2)\cdots))=\pi(jx(1)x(2)\cdots)$. Our notation for $I_{\mathbf{x}(m)}$ and $E_{\mathbf{x}(m)}$ was chosen so that $\pi(C_{\mathbf{x}(m)})=I_{\mathbf{x}(m)}$ and so that $E_{\mathbf{x}(m)}$ is that part of the graph of f lying above the interval $I_{\mathbf{x}(m)}$. It is not difficult to see that π is onto. Furthermore, π is almost 1-1 and only fails to be 1-1 at countably many points in I. This is because the collection of intervals $\{I_i\}$ has the property that

$$\text{int } I_i\cap\text{int } I_j=\varnothing \qquad \text{if} \quad i\neq j,$$

and this implies by induction that for each m the collection $\{I_{\mathbf{x}(m)}\}$ has the same property. Thus if $x\in I_{\mathbf{x}(m)}\cap I_{\mathbf{z}(m)}$, then x is an endpoint of $I_{\mathbf{x}(m)}$ and $I_{\mathbf{z}(m)}$. Hence there are only countably many points of I associated with more than one infinite word (in fact each is associated with exactly two infinite words—just in the same way that $0.09999\cdots = 0.10000\cdots$ in the denary representation of the real numbers). These points are the only ones at which E might fail to be the graph of a function. However, our initial assumptions (3) and (4) ensure that the graph "joins up." In particular, note that if E is a graph constructed in the above manner it must be the graph of a continuous function because E is a closed set.

Definition 1. The *Hölder exponent of f at x* is

$$h_x = \lim_{\varepsilon\to 0} \inf\{(\log|f(x)-f(y)|)/\log|x-y|: y \in B(x;\varepsilon)\}.$$

Lemma 1.

$$h_x = \sup\{\alpha: |f(x)-f(y)| \leq |x-y|^\alpha \textit{ for all } y \textit{ in some neighbourhood of } x\}.$$

Proof. Let γ equal the right-hand side of the latter equation. We first show that $h_x \geq \gamma$. Choose $\delta > 0$ and let $\alpha = h_x - \delta$. In some small ε-neighbourhood of x we have

$$y \in B(x;\varepsilon) \Rightarrow \alpha < (\log|f(x)-f(y)|)/\log|x-y|$$

and so $|f(x)-f(y)| \leq |x-y|^\alpha$. Thus $\alpha = h_x - \delta \leq \gamma$ for any $\delta > 0$, and so $h_x \leq \gamma$. In order to prove that $h_x \leq \gamma$ first choose some $\delta > 0$. In some neighbourhood of x the inequality $|f(x)-f(y)| \leq |x-y|^{\gamma-\delta}$ holds. Taking logarithms gives us that $(\log|f(x)-f(y)|)/\log|x-y| \geq \gamma - \delta$ in some neighbourhood of x and so we must have $h_x \geq \gamma - \delta$ for any $\delta > 0$ which implies that $h_x \geq \gamma$. ∎

We put $h = \inf\{h_x: x \in I\}$ and call h the *Hölder exponent of f*. If there exists a number h_λ such that $h_x = h_\lambda$ for Lebesgue almost all x in I then we call h_λ the *almost everywhere Hölder exponent of f*. An arc C is called an *envelope* of f at x if, in some neighbourhood of x, C is a union of the graphs of g_+ and g_-, where $g_\pm(y) = \pm|x-y|^\alpha + f(x)$. We call α the *exponent* of this envelope. Figure 1 shows an example of an envelope. The graph of f, $\{(x, f(x)): x \in I\}$, is denoted $\mathrm{gr}(f)$.

The following relation between D_B and h is well known. The covering argument used in its proof appears in a paper of Besicovitch and Ursell [BU].

Lemma 2. *Suppose f is a function $f: I \to \mathbf{R}$ with Hölder exponent h. Then*

$$h \leq 2 - D_B(\mathrm{gr}(f)).$$

Proof. Given $\varepsilon > 0$ we have $|f(x)-f(y)| \leq |x-y|^{h-\varepsilon}$ for all $x, y \in I$. We cover the graph of f using squares of side length 2^{-n} in the following way: on each interval $[i/2^{-n}, (i+1)/2^{-n}]$ the graph of f is contained in a rectangle of height $2\cdot 2^{-n(h-\varepsilon)}$ which can be covered by $[2\cdot 2^{-n(h-\varepsilon)}/2^{-n}]+1$ squares of side length 2^{-n}. There are 2^n such rectangles altogether. We can thus cover the whole graph using less than

$$d\cdot 2^n \cdot 2^{n(1-h+\varepsilon)} = d\cdot 2^{n(2-h+\varepsilon)}$$

squares of side length 2^{-n} where d is a constant independent of n. This estimate gives $D_B \leq 2 - h + \varepsilon$. ∎

3. Exponents for a Class of Functions with Linear Scalings

We now consider the special case in which $x_i = i/n$ for $0 \le i < n$ so that $a_i = 1/n = a$ and $v_i^0 = ia$ for $0 \le i < n$. Figures 2 and 3 illustrate two functions constructed using $x_0 = 0$, $x_1 = 0.5$, and $x_2 = 1$. The calculation of $D_B(\mathrm{gr}(f))$ in this case (given as part (i) of Theorem 3 below) has been carried out by Hardin and Massopust [HM]. The map $\pi : \Sigma_n \to I$ here is simply given by

$$x(1)x(2)\cdots \to 0\cdot x(1)x(2)\cdots \qquad (\text{base } n).$$

Theorem 3. *In the above case, if the points (x_i, y_i) are not collinear we have:*

(i) $$D_B(\mathrm{gr}(f)) = 1 + \frac{\log \sum_{i=0}^{n-1} c_i}{-\log a}.$$

(ii) $$h_\lambda = \frac{(1/n)\sum_{i=0}^{n-1} \log c_i}{\log a}.$$

(iii) $$h = \frac{\log(\max_{0\le i<n} c_i)}{\log a}.$$

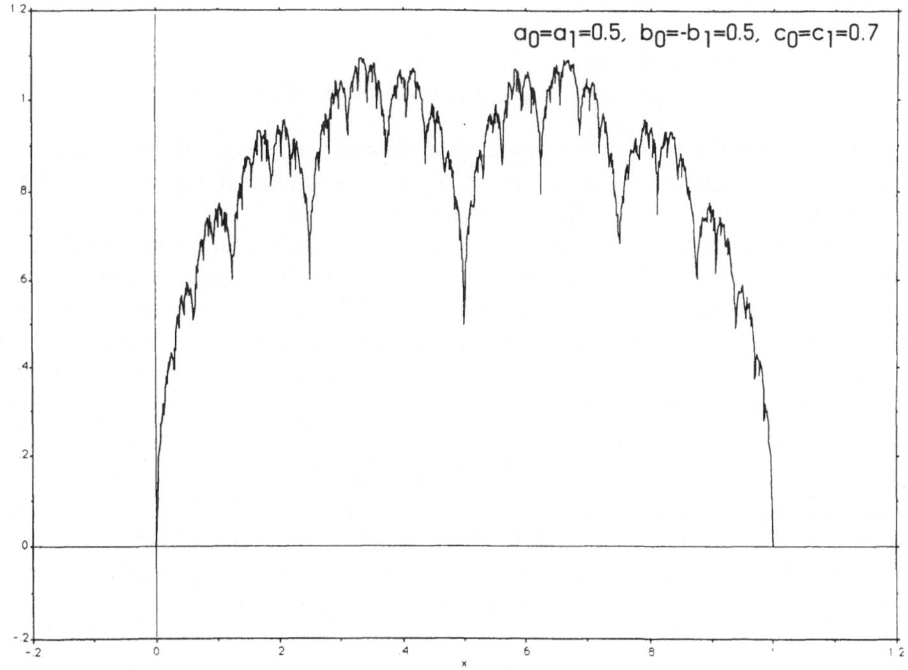

Fig. 2

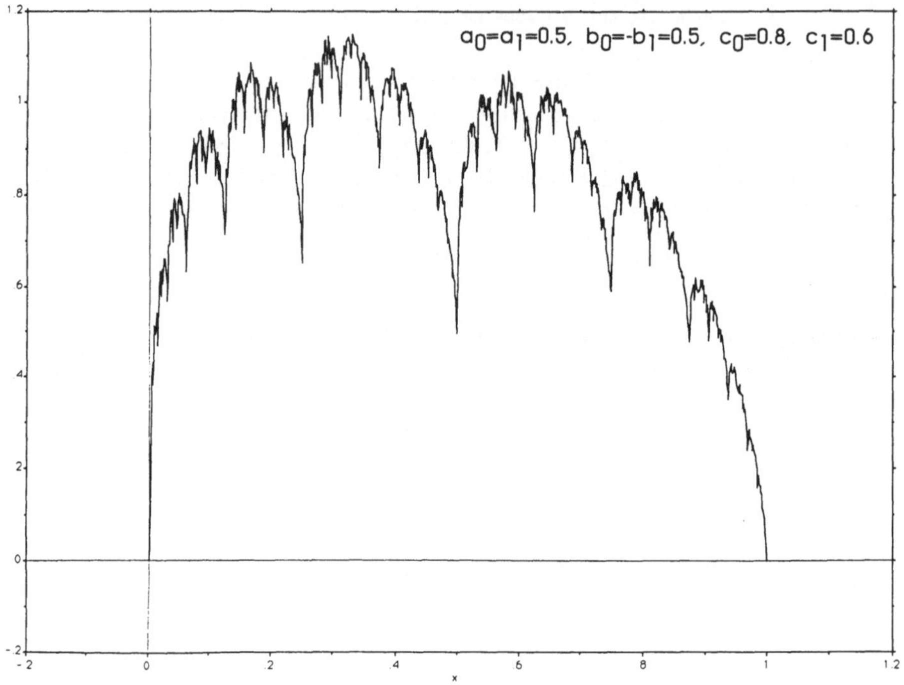

Fig. 3

On the other hand, the following conditions are equivalent:

(a) *The points* (x_i, y_i) *are collinear.*
(b) $\mathrm{gr}(f)$ *is a straight line.*
(c) *The strongly contracting eigenspaces of the maps* φ_i *coincide.*

Proof. We deal first with the degenerate case of (a), (b), and (c). If (a) holds then the line containing the points (x_i, y_i) is invariant under each φ_i, and thus is the strongly contracting eigenspace for each φ_i which implies that (c) holds. If (c) holds then letting F equal the strongly contracting eigenspace of φ_0 intersected with $[0, 1] \times \mathbf{R}$, we have $F = \bigcup_i \varphi_i(F)$ and so by uniqueness of compact nonempty solutions to this equation must equal $E = \mathrm{gr}(f)$. In particular $\mathrm{gr}(f)$ is a straight line, so (c) implies (b). Since $\mathrm{gr}(f)$ contains the points (x_i, y_i), (b) implies (a) is trivial. This shows that (a), (b), and (c) are equivalent. Another way to see that (a) implies (c), which generalizes to the case of the maps φ_i being nonlinear, is as follows. Suppose that (c) does not hold, so that $E = \mathrm{gr}(f)$ is somewhere not equal to the strongly contracting eigenspace of φ_i (say). Now, as $n \to \infty$, $\varphi_i^n(E)$ converges to α_i, the fixed point of φ_i, and converges asymptotically to the weak unstable eigenspace of φ_i at α_i. Since $\varphi_i^n(E) \supseteq E$ for all $n \geq 0$ this shows that E is not contained in the strongly contracting eigenspace of any of the φ_j, and is not a straight line. Note that in a neighbourhood of α_i we can find an arc C of E such that $|C|_H / |C|_W$ is as large as required (where for a subset A of $\mathbf{R}^2$, we define the *height* of A as

$$|A|_H = \sup\{|y - y'|: (x, y), (x', y') \in A\}$$

and the *width* of A as

$$|A|_W = \sup\{|x-x'|: (x,y), (x',y') \in A\}).$$

In order to prove (i), (ii), and (iii) we must first prove the following basic estimates: there exists $d>0$ such that, for all $\mathbf{x} \in \Sigma$ and $m \geq 0$,

$$|E_{\mathbf{x}(m)}|_W = a^m$$

and

$$d^{-1} c_{x(1)} \cdots c_{x(m)} \leq |E_{\mathbf{x}(m)}|_H \leq d c_{x(1)} \cdots c_{x(m)}.$$

The first estimate is clear so we proceed to prove the second and third. Let $b = \max|b_i|$, and $S \subseteq E$. We use the notation $S_{\mathbf{x}(m)} = \varphi_{\mathbf{x}(m)}(S)$. By induction it can be shown that

$$|S_{\mathbf{x}(m)}|_H \leq c_{x(1)} \cdots c_{x(m)} \cdot |S|_H + |S|_W b \sum_{j=1}^{m} a^{m-j+1} \left(\prod_{k=1}^{j-1} c_{\mathbf{x}(k)} \right)$$

and

$$|S_{\mathbf{x}(m)}|_H \geq c_{x(1)} \cdots c_{x(m)} \cdot |S|_H - |S|_W b \sum_{j=1}^{m} a^{m-j+1} \left(\prod_{k=1}^{j-1} c_{\mathbf{x}(k)} \right).$$

This follows straightforwardly from the fact that $\varphi_{x(j-1)}(S_{x(j)\cdots x(m)}) = S_{x(j-1)\cdots x(m)}$, and so

$$|S_{x(j-1)\cdots x(m)}|_H \leq c_{x(j-1)} \cdot |S_{x(j)\cdots x(m)}|_H + b|S_{x(j)\cdots x(m)}|_W$$

and

$$|S_{x(j-1)\cdots x(m)}|_H \geq c_{x(j-1)} \cdot |S_{x(j)\cdots x(m)}|_H - b|S_{x(j)\cdots x(m)}|_W.$$

Rearranging the expressions for $|S_{\mathbf{x}(m)}|_H$ and using $a_i/c_i < 1$, we have

$$|S_{\mathbf{x}(m)}|_H \leq c_{x(1)} \cdots c_{x(m)} \cdot \{|S|_H + |S|_W d_1\}$$

and

$$|S_{\mathbf{x}(m)}|_H \geq c_{x(1)} \cdots c_{x(m)} \cdot \{|S|_H - |S|_W d_1\},$$

where $d_1 > 0$ is a constant independent of m. If we take S in the second inequality to be a small arc in a neighbourhood of α_0 (say) then we can ensure that $|S|_H - |S|_W d_1 > 0$. Taking $S = E$ in the first inequality then proves our claimed estimates with $d = \max\{|E|_H + |E|_W d_1, (|S|_H - |S|_W d_1)^{-1}\}$.

Proof of (i). We are now in a position to calculate D_B. For each m, we cover $\mathrm{gr}(f)$ by squares of the form $[ia^m, (i+1)a^m] \times [ja^m, (j+1)a^m]$ where $m, i, j \in N$. If $N(m)$ is the minimum number of such squares (for fixed m) covering $\mathrm{gr}(f)$ then $[\log N(m)]/[-m \log a] \to D_B$ as $m \to \infty$ if the limit exists [Tr]. Now, using our estimates of $|E_{\mathbf{x}(m)}|_H$ we have

$$d^{-1} \sum_{\mathbf{x}(m)} c_{x(1)} \cdots c_{x(m)} a^{-m} \leq N(m) \leq d \sum_{\mathbf{x}(m)} c_{x(1)} \cdots c_{x(m)} a^{-m}.$$

This gives us

$$d^{-1} a^{-m} (c_0 + \cdots + c_{n-1})^m \leq N(m) \leq d a^{-m} (c_0 + \cdots + c_{n-1})^m$$

and so

$$D_B(\text{gr}(f)) = 1 + \frac{\log \sum_{i=0}^{n-1} c_i}{-\log a}.$$

Proof of (ii). First we show that $h_x \leq H$ for almost all x, where H is the claimed value of h_λ: this is done by estimating the size of the envelope needed at x. Suppose for a contradiction that $h_x > H$, and choose α, β, γ so that $h_x > \alpha > \beta > \gamma > H$. Then in some small neighborhood of x we have $|fx - fy| < |x-y|^\alpha$. For large enough m the m-cylinder $I_{\mathbf{x}(m)}$ containing x is inside this neighborhood. Thus we have

$$(*) \qquad |E_{\mathbf{x}(m)}|_H \leq 2a^{m\alpha} < |E_{\mathbf{x}(m)}|_W^\beta$$

for large enough m. However, from the estimates made in the proof of part (i) we have

$$d^{-1} c_{x(1)} \cdots c_{x(m)} \leq |E_{\mathbf{x}(m)}|_H \leq d c_{x(1)} \cdots c_{x(m)}.$$

Now, for almost all $\mathbf{x}$ (and hence for almost all $x = \pi(\mathbf{x})$),

$$\frac{1}{m} \log(c_{x(1)} \cdots c_{x(m)}) = \frac{1}{m} \sum_{j=1}^{m} \log c_{x(j)} \to \frac{1}{n} \sum_{i=0}^{n-1} \log c_i$$

as $m \to \infty$. Thus for almost all $\mathbf{x}$, if m is large enough

$$(\log|E_{\mathbf{x}(m)}|_H)/(\log|E_{\mathbf{x}(m)}|_W) \leq \gamma$$

and so $|E_{\mathbf{x}(m)}|_H \geq |E_{\mathbf{x}(m)}|_W^\gamma > |E_{\mathbf{x}(m)}|_W^\beta$ which contradicts $(*)$ above. We have therefore shown that $h_x \leq H$ for almost all x.

To see that $h_x \geq H$ for almost all x is more difficult. We denote by P_m^r and P_m^l the m-cylinders to the right and left, respectively, of the one containing x. If $u \neq x$ then u and x are in different m-intervals for high enough m. This implies that if R_m^r and R_m^l are envelopes through x containing $\{(z, f(z)) : z \in P_m^r\}$ and $\{(z, f(z)) : z \in P_m^l\}$, respectively, then $(u, f(u))$ is within R_m^r or R_m^l for some large m. We can thus estimate h_x from below by estimating the exponents of R_m^r and R_m^l. The estimates for the right and left envelopes are similar—below we give the calculation for the right envelope. For convenience we write $R_m = R_m^r$.

There are two problems in making an estimate of the exponent of R_m. The first is that if x is close to the right-hand boundary of $I_{\mathbf{x}(m)}$ the exponent will tend to be larger than if x is close to the left-hand boundary of $I_{\mathbf{x}(m)}$. The second factor is that if the height of P_m is very much larger than the height of $E_{\mathbf{x}(m)}$, the exponent of R_m will not depend greatly on $x(1) \cdots x(m)$. Both of these factors become significant when the sequence $\mathbf{x}$ corresponding to x contains long strings of $(n-1)$'s.

Assume that $x = \pi(\mathbf{x})$ where $\mathbf{x} = x(1) \cdots x(r-k) \cdots x(r) \cdots x(r+m) \cdots$ and that k and m are the smallest positive integers such that $x(r-k), x(r+m) \neq n-1$.

We estimate the exponent of R_r. First note that, by definition of P_r, if $P_r = I_{y(1)\cdots y(r)}$ then $y(1)\cdots y(r-k-1)=x(1)\cdots x(r-k-1)$ with $y(r-k)=x(r-k)+1$, $y(r-k+1)=0$, and $x(r-k+1)=n-1$. The distance of x to P_r is at least a^{r+m} because the $(r+m)$-cylinder $I_{x(1)\cdots x(r+m-1)(n-1)}$ lies between x and P_r. We know from the estimates made above that

$$|E_{\mathbf{x}(r)}|_{\mathrm{H}} \le d\cdot c_{x(1)}\cdots c_{x(r)}$$

and

$$|P_r|_{\mathrm{H}} \le d\cdot c_{y(1)}\cdots c_{y(r)}.$$

The envelope R_r has exponent greater than or equal to

$$\frac{\log(|E_{\mathbf{x}(r)}|_{\mathrm{H}}+|P_r|_{\mathrm{H}})}{(r+m)\log a}$$

which is greater than or equal to

$$\frac{\log c_{x(1)}\cdots c_{x(r-k-1)}}{(r+m)\log a}+\frac{\log(d(c_{x(r-k)}\cdots c_{x(r)}+c_{y(r-k)}\cdots c_{y(r)}))}{(r+m)\log a}.$$

Now, let $j_1(x, r)=m, j_2(x, r)=k$: if we can show that, for almost all $x, j_i(x, r)/r\to 0$ as $r\to\infty$ (for $i=1, 2$), then the proof is complete because our estimate on the exponent of R_r will converge to H for almost all x. To see that $j_i(x, r)/r\to 0$ for almost all x notice that

$$\lambda\{x: j_i(x, r)>m\}\le n^{-m},$$

so that

$$\lambda\{x: j_i(x, r)/r\ge\varepsilon \text{ for some } r\ge N\}\le\sum_{r=N}^{\infty} n^{-r\varepsilon},$$

which goes to zero as $N\to\infty$. Hence $j_i(x, r)/r\to 0$ for almost all x. This concludes the proof of (ii).

Proof of (iii). Suppose that $c_j=\max c_i$. Then for any $\mathbf{x}$ ending with a string of j's we have

$$\frac{\log(|E_{\mathbf{x}(m)}|_{\mathrm{H}})}{\log(|E_{\mathbf{x}(m)}|_{\mathrm{W}})}\to\frac{\log c_j}{\log a}.$$

By the argument given in the proof of (ii), if $x=\pi(\mathbf{x})$ then $h_x<(\log c_j)/(\log a)$. Furthermore, as any finite word in $\{0,\ldots, n-1\}$ can be continued with a string of j's, the set of such x is dense in I. It is also clear that $(\log c_j)/(\log a)$ is a lower bound on h_x for any x. Thus $h=(\log c_j)/(\log a)$. ■

Corollary 4. *For a function f defined as above,*

$$h\le 2-D_{\mathrm{B}}(\mathrm{gr}(f))\le h_\lambda.$$

Both inequalities can be replaced by equalities if and only if the c_i are all equal.

Proof. Recall that $a=1/n$, then we have

$$2-D_{\mathrm{B}}=\frac{\log((1/n)(c_0+\cdots+c_{n-1}))}{\log a}$$

which is less than

$$h_\lambda=\frac{\log((c_0\cdots c_{n-1})^{1/n})}{\log a}. \qquad \blacksquare$$

The functions shown in Figs. 2 and 3 have the same box dimensions ($D_{\mathrm{B}}\approx 1.4854$) but have different exponents. The graph shown in Fig. 2 has $h=h_\lambda\approx 0.5146$, but that in Fig. 3 has $h\approx 0.3219$ and $h_\lambda\approx 0.5294$.

Although above we calculated only the almost everywhere Hölder exponent for Lebesgue measure it is not difficult to see that the same argument works for any other shift-invariant probability measure without an atom at $(0,0,0,\ldots)$ or $(1,1,1,\ldots)$ (because we only need that $j_i(x,r)/r\to 0$ as $r\to\infty$ for almost all x). For example, given $p_0,\ldots,p_{n-1}$ with $\sum p_i=1$ there is a (unique) measure μ with the property that

$$\mu(I_{x(1)\cdots x(m)})=p_{x(1)}\cdots p_{x(m)}.$$

This measure is often called a **p**-balanced measure, a Bernoulli measure, or a Besicovitch measure. As a corollary of the proof of the last theorem we have:

Corollary 5. *If μ is the measure defined above, $h_\mu=(\sum p_i\log c_i)/\log a$.*

The condition $0<a_i<c_i$ that we imposed is required to ensure that the values of b_i do not asymptotically influence $|E_{\mathbf{x}(m)}|_{\mathrm{H}}$ for any $\mathbf{x}$, m. There is an interesting example suggested to me by Benoit Mandelbrot for which $b_i\equiv 0$ so that the condition of $a_i<c_i$ can be relaxed, namely when f is the cumulative distribution function of a Bernoulli measure. To be more precise, let μ be the measure of Corollary 5. For any Borel set B and $0\le i<n$,

$$p_i\cdot\mu(B)=\mu(a\cdot i+a\cdot B)$$

(where $a=1/n$). Hence for the cumulative distribution function for μ, f, we have $\varphi_i(\mathrm{gr}(f))=\mathrm{gr}(f|I_i)$ where $\varphi_i(x,y)=(ax+ai,\,p_iy+\sum_{j<i}p_j)$, and in particular $\mathrm{gr}(f)$ is self-affine (Fig. 4 shows $\mathrm{gr}(f)$ for the $(\frac{1}{3},\frac{2}{3})$ Bernoulli measure). The local Hölder properties of this graph are closely related to the Hausdorff dimension of the set

$$B_p=\{x=0\cdot x_1x_2\cdots\ (\text{base } n)\,|\,(1/m)\cdot\#\{j: 0<j\le m,\ x_j=i\}\to p_i \text{ as } m\to\infty \text{ for } 0\le i<n\}$$

(see [Ca] for a discussion of the Hausdorff dimension of this and other similar sets). For $x\in B_p$ by the argument of the above theorem we have $h_x=(-\sum_i p_i\log p_i)/(\log n)$. Hence $f|B_p$ has Hölder exponent everywhere equal to $(-\sum_i p_i\log p_i)/(\log n)$ and an easy argument shows that

$$\left\{\left(-\sum_i p_i\log p_i\right)\Big/(\log n)\right\}\cdot\dim f(B_p)\le\dim B_p.$$

A similar argument for f^{-1} shows that equality holds above. Now, since f is the cumulative density function for μ and $\mu(B_p)=1$ we have $\dim f(B_p)=1$, which implies that the dimension of B_p is $(-\sum_i p_i\log p_i)/(\log n)$.

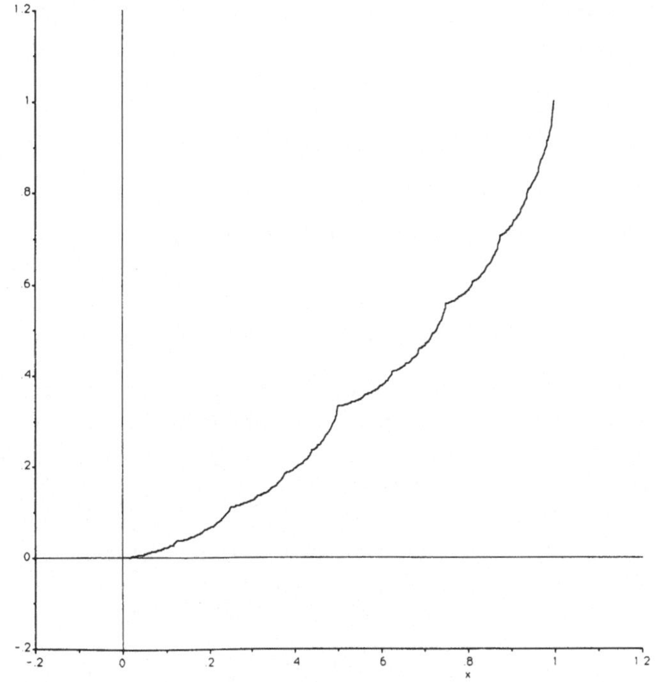

Fig. 4. Cumulative density of the Bernoulli $(p, 1-p)$ measure for $p=\frac{1}{3}$.

A nice way to view Theorem 3 is as follows: the function $f: I \to \mathbf{R}$ is associated with a function $g: \Sigma_n \to \mathbf{R}$, where Σ_n is the set of all infinite words using the symbols $0, \ldots, n-1$ and g is defined by setting $g(\mathbf{x}) = f(\pi(\mathbf{x}))$. The difference between f and g is in the topology of their domains: Σ_n is homeomorphic to the Cantor set whereas $I = \pi(\Sigma_n)$ is obtained under π by sticking together various cylinders of Σ_n (giving rise to points of I with more than one symbol sequence). Now, there is a metric d and a measure Λ on Σ_n with $\pi * (\Lambda) = \lambda$ such that the Λ almost everywhere Hölder exponent of g equals the λ almost everywhere Hölder exponent of f, and so we would like to use π to calculate h_λ. However, π is an unbounded contraction mapping, i.e., there is no $K>0$ such that $|\pi\mathbf{x} - \pi\mathbf{x}'| > K \cdot d(\mathbf{x}, \mathbf{x}')$. This fact makes a quick calculation of h_λ impossible. The second part of the proof of Theorem 3(ii) can be thought of as a check that despite the unbounded contraction of π it still behaves reasonably on most of Σ_n (π has Hölder exponent 1 at Λ almost all points).

4. The Nonlinear Case

First we give the extended definition of a self-affine function. Given a collection of points $(x_0, y_0), (x_1, y_1), \ldots, (x_n, y_n)$, with $x_0 = 0$ and $x_n = 1$, we choose for $i = 0, \ldots, n-1$ a $C^{1+\varepsilon}$ diffeomorphism $\varphi_i: \mathbf{R}^2 \to \mathbf{R}^2$ of the plane with the property that the derivative at each point (x, y) is of the form

$$\begin{pmatrix} a_i(x) & 0 \\ b_i(x) & c_i(x) \end{pmatrix}. \tag{6}$$

(This implies that vertical lines are preserved by φ_i, and so φ_i induces a map ψ_i of I.) We also insist that

$$\varphi_i((x_0, y_0)) = (x_i, y_i),$$

$$\varphi_i((x_n, y_n)) = (x_{i+1}, y_{i+1}),$$

$$0 < a_i(x), \qquad 0 < c_i(x),$$

(7) $$\sup\{a_i(x)/c_i(x) \colon x \in I, 0 \le i < n\} < \beta < 1,$$

and

(8) $$\sup\{c_i(x) \colon x \in I, 0 \le i < n\} < 1.$$

Assumptions (7) and (8) imply that either the family $\{\varphi_i\}$ are contractions or that $\{\varphi_{x(1)} \circ \cdots \circ \varphi_{x(m)}\}$ are contractions on $[0, 1] \times \mathbf{R}$ for some m. Thus again we have a unique compact nonempty set E such that $\bigcup \varphi_i(E) = E$. The main result is that, for this extended class of functions, there exists a number that is the Hölder exponent at almost all points with respect to the Lebesgue measure. First we discuss a degenerate case in which E can be differentiable.

When φ_i is linear there is a submanifold W_i^s (the strongly contracting eigenspace) determined by the property that, for $x \in W_i^s$,

$$|\varphi_i^m(x) - \alpha_i| = o(a_i^m).$$

In the nonlinear case the Stable Manifold Theorem (see, for example, [FHY]) asserts the existence of a differentiable submanifold W_i^s determined by the property that, for all $x \in W_i^s$,

$$|\varphi_i^m(x) - \alpha_i| = o((a_i(\alpha_i) + \varepsilon)^m) \qquad \text{for all} \quad \varepsilon > 0.$$

W_i^s is invariant under φ_i and is both embedded and the graph of a function (due to condition (7)). If $[0, 1] \times \mathbf{R} \cap W_i^s = [0, 1] \times \mathbf{R} \cap W_j^s$ for all $0 \le i, j < n$ then $E = [0, 1] \times \mathbf{R} \cap W_0^s$ satisfies

$$E = \bigcup_i \varphi_i(E)$$

and so must be the self-affine set (in particular E is differentiable). If, for some i and j,

$$[0, 1] \times \mathbf{R} \cap W_i^s \ne [0, 1] \times \mathbf{R} \cap W_j^s,$$

then using a similar argument to the linear case (Theorem 3) we see that E is not differentiable. Hence we have that E is differentiable if and only if the strong stable manifolds W_i^s coincide over $[0, 1]$. In this case

$$E = [0, 1] \times \mathbf{R} \cap W_0^s.$$

From now on we assume that E is not differentiable.

Before giving the extra details necessary to prove our main result we prove two lemmas. The reader should recall that $\pi(\sigma^i \mathbf{x}) \in I_{x(i+1) \cdots x(m)}$, and so the points y_i and z_i in the statement of Lemma 6 (below) are approximately following π of the orbit of $\mathbf{x}$.

Lemma 6. *Suppose that* $y_i, z_i \in I_{x(i+1)\cdots x(m)}$ *for* $i=1,\ldots,m-1$, *and that* $y_m, z_m \in I$. *Suppose also that there are maps* $\varphi_j : I \to \mathbf{R}$ $(j=1,\ldots,m)$, *each of which is Hölder continuous of order* α. *Then there exists* $M>0$ *independent of* $y_0,\ldots,y_{m-1}$, $z_0,\ldots,z_{m-1}$, *and* m *such that*

$$\left|\sum_{j=1}^{m} \varphi_j(y_j) - \sum_{j=1}^{m} \varphi_j(z_j)\right| < M.$$

Proof. We assumed above that $\sup\{a_i(x): i, x\} < \rho < 1$ for some ρ. Thus $|y_j - z_j| < \rho^{j-1}$. But now $|\varphi_j(y_j) - \varphi_j(z_j)| < |y_j - z_j|^{\alpha} < \rho^{\alpha(j-1)}$. Hence

$$\left|\sum_{j=1}^{m} \varphi_j(y_j) - \sum_{j=1}^{m} \varphi_j(z_j)\right| < \sum_{j=1}^{m} \rho^{\alpha(j-1)} < \sum_{j=1}^{\infty} \rho^{\alpha(j-1)} < M. \qquad \blacksquare$$

From now on we let u_i and v_i denote points of $I_{x(i+1)\cdots x(m)}$ $(i=1,\ldots,m)$ at which $a_{x(i)}(y)$ takes respectively its inf and sup values on $I_{x(i+1)\cdots x(m)}$. Similarly, w_i, z_i are points of $I_{x(i+1)\cdots x(m)}$ at which $c_{x(i)}(y)$ takes respectively its inf and sup values on $I_{x(i+1)\cdots x(m)}$.

Lemma 7. *Let* $y_i = \pi(\sigma^i \mathbf{x}) = \psi_{x(i)}^{-1} \circ \cdots \circ \psi_{x(1)}^{-1}(x)$. *There exists* $M>0$ *independent of* m *such that if either* (i) $s_i = w_i$ for $i=1,\ldots,m-1$ *or* (ii) $s_i = z_i$ *for* $i=1,\ldots,m-1$, *then*

$$|\log(c_{x(1)}(s_1)\cdots c_{x(m)}(s_m)) - \log(c_{x(1)}(y_1)\cdots c_{x(m)}(y_m))| < M.$$

Proof. The conditions of Lemma 6 are satisfied because $s_0, y_0 \in I$ and $s_{i-1}, y_{i-1} \in I_{x(i)\cdots x(m)}$ for $i=2,\ldots,m$, and because (by assumption) each of the functions $c_{x(j)}$ is Hölder continuous. ■

A similar statement to that of Lemma 7 is true with the c's replaced by a's. With the notation of Lemma 7, we have $y_i = \pi(\sigma^i\mathbf{x})$. Hence if we define $C:\Sigma_n \to \mathbf{R}$ by $C(\mathbf{x}) = \log c_{x(1)}(x)$, and $A:\Sigma_n \to \mathbf{R}$ by $A(\mathbf{x}) = \log a_{x(1)}(x)$, we have

$$\log(c_{x(1)}(y_1)\cdots c_{x(m)}(y_m)) = \sum_{i=1}^{m} C(\sigma^i\mathbf{x})$$

and

$$\log(a_{x(1)}(y_1)\cdots a_{x(m)}(y_m)) = \sum_{i=1}^{m} A(\sigma^i\mathbf{x}).$$

It is not difficult to see that both A and C are Hölder continuous functions. The Ergodic Theorem implies that if μ is a σ-invariant ergodic measure on Σ_n, then for μ-almost all $\mathbf{x}$,

$$\frac{1}{m}\sum_{i=1}^{m} A(\sigma^i\mathbf{x}) \quad\text{and}\quad \frac{1}{m}\sum_{i=1}^{m} C(\sigma^i\mathbf{x})$$

converge to numbers $C(\mu)$ and $A(\mu)$ for μ-almost all $\mathbf{x}$. We now need to find a σ-invariant ergodic μ that in some way corresponds to the Lebesgue measure on I.

Lemma 8. *There is an ergodic σ-invariant probability measure Λ on Σ_n such that $\pi_*(\Lambda)$ is equivalent to the Lebesgue measure on I.*

Proof. Our assumption that $a_i(x)$ is Hölder continuous implies that A is also Hölder continuous. From Bowen's formula for Hausdorff dimension [Bo1], [Be] we know that the Hausdorff dimension of the invariant set $I = \bigcup \psi_i(I)$ is equal to the unique real r such that the topological pressure of rA is equal to zero. The same result shows that the ergodic equilibrium state measure for rA, Λ, has the property that $\pi_*(\Lambda)$ is equivalent to r-dimensional Hausdorff measure. In this case the Hausdorff dimension of I is 1, so $r = 1$ and $\pi_*(\Lambda)$ is equivalent to Lebesgue measure λ on I.

We need to know the density of Λ. The theory of Gibbs states [Bo2] tells us that there exists $q > 1$ such that, for any r, m, and $\mathbf{z} \in C_{x(r)\cdots x(m)}$,

$$\Lambda(C_{x(r)\cdots x(m)}) \in [q, q^{-1}] \exp\left(\sum_{i=0}^{m-r} A(\sigma^i \mathbf{z})\right). \qquad \blacksquare$$

Theorem 9. *Let f be a generalized self-affine function constructed as above and Λ be the measure constructed in Lemma 8. Then, for λ almost all $x \in I$, $h_x = C(\Lambda)/A(\Lambda)$.*

Proof. The structure of the proof is identical to that of Theorem 3(ii). As before we estimate $|E_{\mathbf{x}(m)}|_{\mathrm{H}}$. Defining $S_{\mathbf{x}(m)}$ as we did in the proof of Theorem 3(ii) we obtain, by induction,

$$\prod_{i=1}^{m} a_{x(i)}(u_i) \le |E_{\mathbf{x}(m)}|_{\mathrm{W}} \le \prod_{i=1}^{m} a_{x(i)}(v_i),$$

$$|S_{\mathbf{x}(m)}|_{\mathrm{H}} \le |S|_{\mathrm{H}} \prod_{i=1}^{m} c_{x(i)}(z_i) + |S|_{\mathrm{W}} b \sum_{i=1}^{m} \prod_{k=1}^{j-1} c_{x(k)}(z_k) \prod_{k=j+1}^{m} a_{x(k)}(v_k),$$

and

$$|S_{\mathbf{x}(m)}|_{\mathrm{H}} \ge |S|_{\mathrm{H}} \prod_{i=1}^{m} c_{x(i)}(z_i) - |S|_{\mathrm{W}} b \sum_{i=1}^{m} \prod_{k=1}^{j-1} c_{x(k)}(z_k) \prod_{k=j+1}^{m} a_{x(k)}(v_k).$$

Following the argument of Theorem 3(ii) we find a $d > 0$ such that

$$d \cdot c_{x(1)}(z_1) \cdots c_{x(m)}(z_m) \le |E_{\mathbf{x}(m)}|_{\mathrm{H}} \le d \cdot c_{x(1)}(z_1) \cdots c_{x(m)}(z_m).$$

We can now apply Lemmas 6 and 7 to show that, for Λ almost all $\mathbf{x}$,

$$(1/m) \log|E_{\mathbf{x}(m)}|_{\mathrm{W}} \to A(\Lambda) \quad \text{and} \quad (1/m) \log|E_{\mathbf{x}(m)}|_{\mathrm{H}} \to C(\Lambda)$$

as $m \to \infty$. The first part of the proof of Theorem 3(ii) goes through to show that $h_x \le H = C(\Lambda)/A(\Lambda)$ for Lebesgue almost all x. The proof that $h_x \ge H$ for almost all x is also similar to the case dealt with previously. We use the notation established before and estimate the exponent of R_r. The distance between x and the next r-cylinder, P_r, is more than $|E_{x(1)\cdots x(r+m-1)(n-1)}|_{\mathrm{W}}$ which is bounded below by

$$a_{x(1)}(u_1) \cdots a_{x(r)}(u_r) \cdot \zeta^m,$$

where $0<\zeta\leq\inf\{a_i(y): y\in I, 0\leq i<n\}$. The envelope R_r thus has exponent greater than or equal to

$$\frac{\log(c_{x(1)}(z_1)\cdots c_{x(r-k-1)}(z_{r-k-1}))}{m\log\zeta+\log(a_{x(1)}(u_1)\cdots a_{x(r)}(u_r))}+\frac{\log(2k\tau)}{(r+m)\log\zeta},$$

where $\tau\geq\{c_i(y): y\in I, 0\leq i<n\}$. Writing $j_1(\mathbf{x}, r)=m$ and $j_2(\mathbf{x}, r)=k$, if we can show that $j_i(\mathbf{x}, r)/r\to 0$ as $r\to\infty$ for almost all $\mathbf{x}$ ($i=1,2$) we are finished because (using Lemma 7) the estimate for the exponent of R_r will have the same limit as

$$\frac{(1/r)\sum_{i=0}^{r-1} C(\sigma^i\mathbf{x})}{(1/r)\sum_{i=0}^{r-1} A(\sigma^i\mathbf{x})}$$

which is $C(\Lambda)/A(\Lambda)$ as claimed.

In order to show $j_i(\mathbf{x}, r)/r\to\infty$ we have to estimate the probability of a block of $(n-1)$'s of length j. From the estimate for the measure of a cylinder set given at the end of Lemma 7, we see that this probability is less than ξ^j, where $1>\xi\geq\sup\{a_i(y): y\in I, 0\leq i<n\}$. Hence

$$\Lambda\{\mathbf{x}: j_i(\mathbf{x}, r)/r\geq\varepsilon \text{ for some } r\geq N\}\leq\sum_{r=N}^{\infty}\zeta^{-r\varepsilon},$$

which converges to zero as $N\to\infty$. This shows that $j_i(\mathbf{x}, r)/r\to 0$ for Λ-almost all $\mathbf{x}$ and hence also for λ-almost all x and completes the proof. ∎

Corollary 10. *Suppose that each of the functions $a_i(x)$ and $c_i(x)$ are constant, $a_i(x)=a_i$ and $c_i(x)=c_i$, then $h_\lambda=(\sum a_i\log c_i)/(\sum a_i\log a_i)$.*

Proof. In this case Λ is the Bernoulli measure given $(p_0,\ldots,p_{n-1})=(a_0,\ldots,a_{n-1})$. ∎

It is not true in general that $h_\lambda\geq 2-D_B$. Consider a collection of φ_i which are linear maps but with a_i not necessarily equal to a_j if $i\neq j$. D. Hardin has shown [Ha] that D_B is given in this case as the unique D with the property that $\sum c_i a_i^{D-1}=1$. If we take $c_0=c_1=0.7$ and $a_0=0.6$, $a_1=0.4$ we have $D_B\approx 1.48$ and $h_\lambda\approx 0.5<2-D_B$.

Acknowledgments. I would like to thank Benoit Mandelbrot who made it possible for me to stay in Harvard and Michael Barnsley who enabled me to visit the Georgia Institute of Technology in April 1986. The ideas contained in this paper came out of that visit. I would also like to thank Colin Sparrow for his helpful comments and suggestions.

References

[Ba] M. F. BARNSLEY (1985): Fractal Functions and Interpolation. Atlanta: Georgia Institute of Technology Preprints.

[BH] M. F. BARNSLEY, A. N. HARRINGTON (1985): The Calculus of Fractal Interpolation Functions. Atlanta: Georgia Institute of Technology Preprints.

[Be] T. J. BEDFORD (1986): *Dimension and dynamics for fractal recurrent sets.* J. London Math. Soc. (2), **33**:89–100.

[BU] A. S. BESICOVITCH, H. D. URSELL (1937): *Sets of fractional dimensions, V: On dimensional numbers of some continuous curves.* J. London Math. Soc., **12**:18–25.

[Bo1] R. BOWEN (1975); Equilibrium States and the Ergodic Theory of Anosov Diffeomorphisms. Lecture Notes in Mathematics, Vol. 470. Berlin: Springer-Verlag.

[Bo2] R. BOWEN (1979): Hausdorff Dimension of Quasi-Circles. Publications Mathématiques, Vol. 50. Paris: Institut des Hautes Etudes Scientifiques, pp. 11–25.

[Ca] H. CAJAR (1981): Billingsley Dimension in Probability Spaces. Lecture Notes in Mathematics, Vol. 892. Berlin: Springer-Verlag.

[FHY] A. FATHI, M. R. HERMAN, J. C. YOCCOZ (1981): *A proof of Pesin's stable manifold theorem.* In: Geometric Dynamics (J. Palis Jr., ed.). Lecture Notes in Mathematics, Vol. 1007. Berlin, Springer-Verlag.

[Ha] D. P. HARDIN (1986): Personal communication.

[HM] D. P. HARDIN, P. R. MASSOPUST (1986): *The capacity for a class of functions.* Comm. Math. Phys., **105**:455–460.

[Hu] J. E. HUTCHINSON (1981): *Fractals and self-similarity.* Indiana Univ. Math. J., **30**:713–747.

[Tr] C. TRICOT: *Two definitions of fractional dimension.* Math. Proc. Cambridge Philos. Soc., **91**:57–74.

T. Bedford[1]
King's College Research Centre
King's College
Cambridge CB2 1ST
England

[1] Current address: Department of Mathematics and Informatics, Delft University of Technology, P.O. Box 356, Delft, The Netherlands.

Constr. Approx. (1989) 5: 49–68

CONSTRUCTIVE APPROXIMATION

Symmetric Iterative Interpolation Processes

Gilles Deslauriers and Serge Dubuc

Abstract. Using a base b and an even number of knots, we define a symmetric iterative interpolation process. The main properties of this process come from an associated function F. The basic functional equation for F is that $F(t/b) = \sum_n F(n/b)F(t-n)$. We prove that F is a continuous positive definite function. We find almost precisely in which Lipschitz classes derivatives of F belong. If a function y is defined only on integers, this process extends y continuously to the real axis as $y(t) = \sum_n y(n)F(t-n)$. Error bounds for this iterative interpolation are given.

1. Introduction

We introduce a family of interpolation processes. This family uses two positive integral parameters: b, a base, and $2N$, an even number of moving nodes. Given a function y which is defined on integers, we extend this function to all integral multiples of $1/b$: if r is an integer between 0 and b, and if n is an integer, $y(n+r/b)$ is defined as the value $p(n+r/b)$ where p is the Lagrange polynomial of degree smaller than $2N$ such that $p(k) = y(k)$ for every integer k of $[n-N+1, n+N]$. This construction can be iterated. In this way, an extension $y(t)$ is found for any rational number t whose denominator is an integral power of b.

Our main concern is to prove that the extension $y(t)$ is uniformly continuous on any finite interval whatever the base b, the even number $2N$ of nodes, and the initial values $y(n)$. For each base b and each number N, there is a fundamental function $F(t)$, the iterative interpolation of the sequence $F(n)$, which is equal to 1 at $n=0$ and equal to 0 elsewhere. The main properties of the interpolation process come from its fundamental function. The basic functional equation for F is

$$F(t/b) = \sum_n F(n/b)F(t-n).$$

Date received: January 19, 1987. Date revised: November 9, 1987. Communicated by Michael F. Barnsley.

AMS classification: 65D05, 41A05, 65F15, 15A18, 42A38, 26A16, 65D10, 42A05, 65Q05, 42A85, 47B37.

Key word and phrases: Interpolation, Eigenvalue, Eigenvector, Fourier transform, Lipschitz classes, Curve fitting, Trigonometric polynomials, Recurrence relations, Factorization, Operators in sequence spaces.

Analysis of F is carried through to the study of two mathematical companions of the process: the trigonometric polynomial

$P(\theta)=\sum_k F(k/b)\, e^{ik\theta}$ and the infinite matrix $A=(F(k/b-j))_{-\infty<k<\infty,-\infty<j<\infty}$.

We see that for a symmetric iterative interpolation process which uses Lagrange polynomials, $P(\theta)$ is always nonnegative and is divisible by another trigonometric polynomial: $\sin^{2N}(b\theta/2)/\sin^{2N}(\theta/2)$. The Lipschitz classes to which derivatives of F belong are related to the eigenvalues of matrix A. We compute derivatives of F using linear equations and interpolation theory. We derive error bounds for iterative interpolation when applied to a specific function.

2. Iterative Interpolation with a Base b

In [3] we described a new method of interpolation. The starting point is a function $y(n)$ which is defined on integers. We would like to extend this function to the whole real axis if possible. First, an integer larger than one, b, is used; this will be the base of the interpolation process. $\mathbf{B}_n$ is the set of b-adic rational numbers m/b^n where m is integral, $n=0,1,2,\ldots$, and $y(t)$ is already defined over $\mathbf{B}_0$. The basic idea is to extend the function y to $\mathbf{B}_1, \mathbf{B}_2, \mathbf{B}_3, \ldots$ by induction. In order to extend y from $\mathbf{B}_0$ to $\mathbf{B}_1$, a sequence of parameters $\{p_k\}_{k=-\infty,\infty}$ are used. Some conditions are assumed about this sequence. First, $p_0=1$ and $p_k=0$ for any other integral multiple of b: $k=\pm b, \pm 2b, \pm 3b, \ldots$. In order to simplify the discussion, we also assume that only a finite number of the parameters p_k are different from zero. With these hypotheses, the extension of y is $y(j/b)=\sum_k p_{j-kb}y(k)$. The extension to $\mathbf{B}_2, \mathbf{B}_3, \ldots$ can be carried through in the same way: if j is an integer, we set

$$y(j/b^{n+1})=\sum_k p_{j-kb}\,y(k/b^n). \tag{2.1}$$

The properties of the parameters allow a consistent extension with no restriction on the original sequence $y(n)$. This extension of the sequence $y(n)$ to the set of b-adic rational numbers is simply written as $y(t)$ and is called the iterative interpolation with base b.

Any sequence $y(n)$ can be extended to the set $\mathbf{B}$ of b-adic numbers according to scheme (2.1). This interpolation process satisfies some simple properties. First, it is a *linear process.* Another basic property is the *translation property of the interpolation*: if $y(n)$ is a given sequence with $y(t)$ as interpolation and if m is a given integer, the interpolation of the sequence $y(n+m)$ is the function $y(t+m)$. The following property is called the *homogeneity of the process of interpolation*: if $y(t)$ is the interpolation of the sequence $y(n)$ defined over $\mathbf{B}_0$ and if h is a negative integral power of b, then the interpolation of the sequence $y(nh)$ is the function $y(th)$. These three properties can be proven rather easily.

3. The Fundamental Function of an Iterative Process

Let us start with the following sequence: $y(0)=1$ and $y(n)=0$ when n is any other integer. Using scheme (2.1), this sequence can be extended to $\mathbf{B}$, the b-adic numbers. Let us denote by $F(t)$ this extension, F is called the *fundamental function* of the interpolation process. There is a very close connection between the weights p_k and the values of F on $\mathbf{B}_1$: $p_k = F(k/b)$. This is a direct consequence from (2.1). Let us look for places where F vanishes.

Lemma 3.1. *If N is the smallest index k for which $p_k \neq 0$ and if N^* is the largest index k for which $p_k \neq 0$, then the fundamental function of the interpolation process vanishes outside* $(N/(b-1), N^*/(b-1))$.

Proof. We set $t_0 = 0$ and t_n is defined through the recurrence $t_n = t_{n-1} + Nb^{-n}$. The sequence t_n is decreasing and converges to $N/(b-1)$. As is easily verified from relation (2.1), the restriction of F to $\mathbf{B}_n$ vanishes on $(-\infty, t_n)$, where $n = 1, 2, 3, \ldots$. For any b-adic number $t \leq N/(b-1)$, $F(t) = 0$. A similar argument shows that $F(t^*)$ is also zero if $t^* \geq N^*/(b-1)$. ■

As in [3], the next two theorems can be proved.

Theorem 3.2. *If $y(n)$ is a function defined on integers, the extension $y(t)$ to b-adic numbers according to scheme* (2.1) *is $\sum_n y(n)F(t-n)$ which is basically a finite series.*

Theorem 3.3. *The fundamental function F fulfils the functional equation*

$$F(t/b^m) = \sum_n F(n/b^m)F(t-n). \tag{3.1}$$

4. Some Examples of Iterative Interpolation

We describe a class of iterative interpolation connected with polynomial interpolation. If $\mathbf{S}$ is a given finite set of integers and if $y(n)$ is a function already defined over the set of integers, then it is possible to extend y to the set $\mathbf{B}_1$ by using polynomial interpolation over translates of $\mathbf{S}$ in the following way. If $t = n + r/b$ (where integer r is between 1 and $b-1$) is a point of $\mathbf{B}_1$ where the extension has to be done and if $P(x)$ is the interpolating polynomial such that $P(n+k) = y(n+k)$ for every integer k of $\mathbf{S}$, then we set $y(t) = P(t)$. If $\{L_k\}_{k\in\mathbf{S}}$ areLagrange polynomials when the integers of $\mathbf{S}$ are used as nodal points, then the extension is

$$y(n+r/b) = \sum_{k\in\mathbf{S}} L_k(r/b)y(n+k). \tag{4.1}$$

The fundamental function F has the following basic values: if $-n$ belongs to $\mathbf{S}$ and if r is an integer between 1 and $b-1$, then

(4.2) $F(n+r/b) = L_{-n}(r/b);\qquad F(0) = 1$ and, for all other values of $\mathbf{B}_1$,

$$F = 0.$$

Definition. The interpolation process (4.1) is called a *Lagrange iterative interpolation*. The main property of this interpolation process is provided by the following statement whose proof is easily obtained.

Lemma 4.1. *For any polynomial P of degree smaller than the number of points of* **S**, *the Lagrange iterative interpolation of the sequence $y(n)=P(n)$ is precisely the function $y(t)=P(t)$.*

When the nodal points are $\mathbf{S}=\{-N+1, -N+2, \ldots, N-1, N\}$ for some positive integer N, we speak of *symmetric* interpolation; indeed, symmetric data $y(n)$ produces symmetric interpolation $y(t)$, the fundamental function $F(t)$ is itself symmetric. For example if $\mathbf{S}=\{-1, 0, 1, 2\}$, we get four Lagrange polynomials:

$$L_{-1}(x)=-x(1-x)(2-x)/6, \qquad L_0(x)=(x+1)(1-x)(2-x)/2,$$

$$L_1(x)=(x+1)x(2-x)/2, \qquad L_2(x)=-(x+1)x(1-x)/6.$$

When base b is 2 or 3, the main values of function F are respectively:

t	1/2	3/2
$F(t)$	9/16	−1/16

t	1/3	2/3	4/3	5/3
$F(t)$	20/27	10/27	−5/81	−4/81

If $\mathbf{S}=\{-2, -1, 0, 1, 2, 3\}$, the six Lagrange polynomials are:

$$L_{-2}(x)=(x+1)x(1-x)(2-x)(3-x)/120,$$

$$L_{-1}(x)=-(x+2)x(1-x)(2-x)(3-x)/24,$$

$$L_0(x)=(x+2)(x+1)(1-x)(2-x)(3-x)/12,$$

$$L_1(x)=(x+2)(x+1)x(2-x)(3-x)/12,$$

$$L_2(x)=-(x+2)(x+1)x(1-x)(3-x)/24,$$

$$L_3(x)=(x+2)(x+1)x(1-x)(2-x)/120$$

When the base is 2 or 3, the basic values of F are respectively:

t	1/2	3/2	5/2
$F(t)$	75/128	−25/256	3/256

t	1/3	2/3	4/3	5/3	7/3	8/3
$F(t)$	560/729	280/729	−70/729	−56/729	8/729	7/729

Definition. In order to avoid a long expression, we speak of the *interpolation process of type* (b, N) instead of the Lagrange iterative symmetric interpolation process when the base is b and when the set of nodal points **S** is $\{-N+1, -N+2, \ldots, N-1, N\}$.

5. Fourier Transform of an Iterative Interpolation

We recall some results we have already shown [3] about general iterative interpolation. These results will help us later in the study of properties of the fundamental interpolating function. A sequence of trigonometric polynomials is associated to the fundamental function $F(t)$: $P_n(\theta)=\sum_k F(k/b^n)\, e^{ik\theta}$. If we set $P(\theta)=P_1(\theta)$, we get a recurrence from formula (3.1): $P_{n+1}(\theta)=P_n(b\theta)P(\theta)$. Hence

$$P_n(\theta)=\prod_{0\le k\le n-1} P(b^k\theta). \tag{5.1}$$

Definition. The trigonometric polynomial $P(\theta)$ is said to be the *characteristic polynomial* of the interpolation scheme (2.1).

Under very weak assumption, there is always a Schwartz distribution accompanying the function F. This assumption is that $\sum_k F(k/b)=b$. In that case, if, for any C^∞-function φ, $T_n(\varphi)=\sum_m F(m/b^n)\varphi(m/b^n)/b^n$ (T_n is a linear combination of Dirac masses), then, in [3], we have proven that this sequence of distributions converges weakly to a distribution T.

Theorem 5.1. *If $\sum_k F(k/b)=b$, then the sequence of distributions T_n converges weakly toward a distribution T. If $G(\xi)$ is the Fourier transform of T, then $G(\xi)=\prod_{k\ge 1}[P(-\xi/b^k)/b]$. $G(0)=1$ and the following functional equation is satisfied*:

$$G(b\xi)=G(\xi)\left(\sum_k F(k/b)\, e^{-ik\xi}\right)\Big/ b. \tag{5.2}$$

The Fourier transform of T_n is in fact $G_n(\xi)=P_n(-\xi/b^n)/b^n=\prod_{1\le k\le n}[P(-\xi/b^k)/b]$. The sequence $G_n(\xi)$ converges pointwise to $G(\xi)$.

Definition. The function G is called the *Fourier transform* of the interpolation process.

6. Continuity of the Fundamental Function

We consider an interpolation process of type (b, N). We study the Lagrange iterative interpolation when nodal points are given as $\{-N+1, -N+2, \ldots, 0, 1, \ldots, N-1, N\}$. We prove that its fundamental function is continuous. The main step is to show that the characteristic polynomial of the process is everywhere nonnegative.

Theorem 6.1. *If $P(\theta)$ is the characteristic function of an interpolation process of type (b, N), then any integral multiple of $2\pi/b$ which is not an integral multiple of 2π is a root of P with multiplicity at least equal to $2N$. Moreover, the first $2N-1$ derivatives of P at $\theta=0$ vanish.*

Proof. By definition, $P(\theta)=\sum_k F(k/b)\,e^{ik\theta}$. If $\theta=2\pi r/b$ (where r is a positive integer smaller than b) and if $w=e^{i2\pi r/b}$, then

$$P(\theta)=\sum_k F(k/b)w^k=\sum_{0\le s\le b-1}\sum_n F(n+s/b)w^{nb+s},$$

$$P(\theta)=\sum_{0\le s\le b-1}\sum_n F(n+s/b)w^s.$$

Since $1=\sum_n F(x-n)$ for every x (application of Lemma 4.1 and Theorem 3.2 with the polynomial equal to the constant 1),

$$P(\theta)=\sum_{0\le s\le b-1} w^s=(w^b-1)/(w-1)=0.$$

If $\theta=2\pi r/b$ (with $0\le r<b$), if $w=e^{i2\pi r/b}$, if m is a positive integer smaller than $2N$, and if $P_m=P^{(m)}(\theta)/i^m$, then

$$P_m=\sum_k k^m F(k/b)w^k=\sum_{0\le s\le b-1}\sum_n (n+s/b)^m F(n+s/b)w^{nb+s},$$

$$P_m=\sum_{0\le s\le b-1}\sum_n (n+s/b)^m F(n+s/b)w^s,$$

$$P_m=(-1)^m\sum_{0\le s\le b-1}\sum_n (n-s/b)^m F(s/b-n)w^s.$$

Since $(x-c)^m=\sum_n (n-c)^m F(x-n)$ for every x and for every c whenever $m<N$ (as shown by Lemma 4.1 and Theorem 3.2),

$$P_m=\sum_{0\le s\le b-1}(s/b-s/b)^m w^s=0. \qquad \blacksquare$$

Theorem 6.2. *The characteristic polynomial of an interpolation process of type* (b, N) *is everywhere nonnegative.*

Proof. The degree of the characteristic polynomial $P(\theta)$ is $bN-1$. The symmetry of $F(t)$ makes sure that $P(\theta)$ takes real values. $P(\theta)$ vanishes at the following points: $2\pi k/b$, $k=1,2,\ldots,b-1$; the multiplicity of each of these roots is at least $2N$. Moreover, the first $2N-1$ successive derivatives of P' vanish at zero. Let us look at the critical points of P, the roots of the polynomial P'. $P'(\theta)$ vanishes at the following points: $2\pi k/b$, $k=0,1,\ldots,b-1$; the multiplicity of each of these roots is at least $2N-1$. According to Rolle's theorem, in each interval $(2\pi k/b, 2\pi(k+1)/b)$, $k=1,2,\ldots,b-2$, $P'(\theta)$ is zero for at least one point θ. The number of roots of P' is then at least $2Nb-2$, the degree of P' as a trigonometric polynomial is $bN-1$. Since the number of roots of a trigonometric polynomial cannot exceed the double of its degree, in each interval $(2\pi k/b, 2\pi(k+1)/b)$, $k=1,2,\ldots,b-2$, $P'(\theta)$ is zero at exactly one point and there is no root of P' in $(0, 2\pi/b)$ and in $(-2\pi/b, 0)$. Since $P(0)=b$ and $P(2\pi/b)=0$, P' is negative on $(0, 2\pi/b)$. The derivative $P^{(2N)}(2\pi/b)$ is positive, otherwise $P'(\theta)$ would have too many roots. When the base b is larger than 2, it can be seen that $P^{(2N)}(4\pi/b)$ is positive; first, $P^{(2N)}(4\pi/b)\ne 0$ according to the number of roots of P'; if $P^{(2N)}(4\pi/b)<0$, P would be positive around $2\pi/b$, negative around $4\pi/b$, and there would be two critical points for P in

$(2\pi/b, 4\pi/b)$, but this is impossible. This same argument can be carried through for each interval $(2\pi k/b, 2\pi(k+1)/b)$, $k=2,3,\ldots,b-2$. It follows that $P(\theta)$ never take negative values. ■

Remark. As shown in the previous proof, $P^{(2N)}$ is positive at any integral multiple of $2\pi/b$ which is not an integral multiple of 2π. So any integral multiple of $2\pi/b$ which is not an integral multiple of 2π is a root of P with multiplicity *precisely* equal to $2N$.

Theorem 6.3. *For any base b, for any even positive integer N, the Fourier transform $G(\xi)$ of the interpolation process of type (b, N) is integrable.*

Proof. As already noted (Theorem 5.1), the Fourier transform $G(\xi)$ of the interpolation process is $\prod_{k\geq 1}[P(-\xi/b^k)/b]$. According to the previous theorem, $G(\xi)$ is nonnegative. If J_n is the integral of G over $(-\pi b^n, \pi b^n)$, then $J_n = \int_{(-\pi,\pi)} G(\xi)\prod_{0\leq k\leq n-1} P(-\xi b^k)\, d\xi$, since $G(b^n\xi)$ is equal to $G(\xi)\prod_{0\leq k\leq n-1}[P(-\xi b^k)/b]$. If M is the maximal value of $G(\xi)$ on $[-\pi, \pi]$, then $J_n \leq M \int_{(-\pi,\pi)} \prod_{0\leq k\leq n-1} P(-\xi b^k)\, d\xi$. We already know that the Fourier series of $\prod_{0\leq k\leq n-1} P(\xi b^k)$ is $\sum_k F(k/b^n)\, e^{ik\xi}$. So $\int_{(-\pi,\pi)}\prod_{0\leq k<n-1} P(-\xi b^k)\, d\xi = F(0) = 1$. Each integral J_n is bounded by M. $G(\xi)$ is then integrable. ■

Theorem 6.4. *The fundamental function $F(t)$ of an interpolation process of type (b, N) has a unique continuous extension to the real axis. The Fourier transform of that extension is $G(\xi)$.*

Proof. The Fourier transform $G(\xi)$ of the interpolation process is integrable. $G(\xi)$ is then the Fourier transform of a continuous function $\Phi(t)$: in fact, $\Phi(t) = 1/(2\pi)\int G(\xi)\, e^{it\xi}\, d\xi$. From the functional relation (5.2), $G(b\xi) = G(\xi)P(-\xi)/b$, it follows that $\Phi(t/b) = \sum_k F(k/b)\Phi(t-k)$. The interpolation process applied to the sequence $\Phi(n)$ reproduces $\Phi(t)$ for any b-adic number t. According to Theorem 3.2, for any b-adic number t, $\Phi(t) = \sum_k \Phi(k)F(t-k)$. If we introduce the sequence of trigonometric polynomials $\Pi_n(\theta) = \sum_k \Phi(k/b^n)\, e^{ik\theta}$ and if $Q(\theta)$ is the trigonometric polynomial $\sum_k \Phi(k)\, e^{ik\theta}$, then $\Pi_n(\theta) = Q(b^n\theta)P_n(\theta)$, as is seen after a short computation. The sequence of Fourier transforms $\Pi_n(\xi/b^n)/b^n$ converges to $Q(\xi)G(\xi)$. Since $\Pi_n(\xi/b^n)/b^n$ is the Fourier transform of the following measure μ_n, a linear combination of Dirac masses, $\sum_k b^{-n}\Phi(k/b^n)\delta_{k/b}n$, and since this sequence of measures converges weakly to $\Phi(t)$, we get $Q(\xi)G(\xi) = G(\xi)$. From this relation $Q(\xi) = 1$, $\Phi(0) = 1$, and $\Phi(n) = 0$ for any integer other than zero. $\Phi(t) = F(t)$ for any b-adic number. ■

7. The Order of Regularity of the Fundamental Function

We already know that the fundamental function F of any Lagrange iterative symmetric interpolation process is continuous. Is this function differentiable? What happens for derivatives of higher order? In this section we show some results about the order of regularity of F. The two basic tools of this study are factorization of the characteristic function $P(\theta)$ and finding real values α for which $\int|\xi|^\alpha G(\xi)\,d\xi$ is finite (where G is the Fourier transform of the iterative process). We recall the definition of a *Lipschitz function of order* α; f is such a function if there is a number L such that for any two numbers x_1 and x_2, $|f(x_1)-f(x_2)|\le L|x_1-x_2|^\alpha$. The class of all Lipschitz functions of order α is customarily noted as Lip α. Some authors prefer to use the expression Hölder functions instead of Lipschitz functions. The following lemma makes a connection between the finiteness of a Fourier transform multiplied by a power and Lipschitz functions. This lemma should be known in the mathematical literature, but we were unable to find a reference about it.

Lemma 7.1. *Let $f(t)$ be a function whose Fourier transform is $g(\xi)$. We assume that $|\xi|^{m+\alpha}g(\xi)$ is integrable where m is a nonnegative integer and α is a real number of* $[0, 1]$. *If $\alpha=0$, then $f^{(m)}$ is continuous. If $\alpha\neq 0$, then $f^{(m)}$ belongs to* Lip α.

Proof. (a) That $f^{(m)}$ is continuous when $|\xi|^m g(\xi)$ is integrable is a classical result as proven in Bochner and Chandrasekharan [1].

(b) Let us assume that $\int|\xi|^\alpha|g(\xi)|<\infty$ where $\alpha\in(0,1)$. If t_1 and t_2 are two numbers such that $|t_1-t_2|<\delta$, then

$$2\pi|f(t_1)-f(t_2)|\le\int|e^{i\xi t_1}-e^{i\xi t_2}||g(\xi)|\,d\xi\le\int\min(|\xi|\delta,2)|g(\xi)|\,d\xi.$$

If $\alpha\in[0,1]$, then, for every positive x, $\min(x,1)\le x^\alpha$. So

$$2\pi|f(t_1)-f(t_2)|\le 2\int(|\xi|\delta/2)^\alpha|g(\xi)|\,d\xi.$$

$|f(t_1)-f(t_2)|\le\int|\xi|^\alpha|g(\xi)|\,d\xi(\delta/2)^\alpha/\pi$ and then f belongs to Lip α.

(c) The general case follows rather easily from (a) and (b). ■

Theorem 7.2. *If $P(\theta)$ is the characteristic polynomial of the interpolation process of type (b, N), then there is a trigonometric polynomial $S(\theta)$ of degree $N-1$ such that*

$$P(\theta)=S(\theta)[\sin(b\theta/2)/\sin(\theta/2)]^{2N}. \tag{7.1}$$

Proof. Let us introduce the following polynomial: $\Pi(w)=\sum_{|k|<bN}F(k/b)w^{k+2bN-1}$. From Theorem 6.1, each point $2\pi k/b$ $(k=1,2,\dots,b-1)$ is a root of multiplicity $2N$ of the equation $P(\theta)=0$, each b-root ω of unity other than 1 is a root of multiplicity $2N$ of $\Pi(w)$. $(w-\omega)^{2N}$ is a factor of $\Pi(w)$. If $\omega_1,\omega_2,\dots,\omega_{b-1}$ are these b-roots of unity other than 1, then $\prod_{0\le k\le b}(w-\omega_k)^{2N}$ is a factor of $\Pi(w)$. $\prod_{0\le k\le b}(w-\omega_k)=(w^b-1)/(w-1)$. $\Pi(w)[(w-1)/(w^b-1)]^{2N}$ is a polynomial $\Sigma(w)$ of degree $2N-2$. The following formulas hold:

$$P(\theta)=e^{-i\theta(bN-1)}\Pi(e^{i\theta})=e^{-i\theta(bN-1)}[(e^{ib\theta}-1)/(e^{i\theta}-1)]^{2N}\Sigma(e^{i\theta}),$$

$$P(\theta)=[\sin(\theta b/2)/\sin(\theta/2)]^{2N}e^{-i\theta(N-1)}\Sigma(e^{i\theta}).$$

Since $e^{-i\theta(N-1)}\Sigma(e^{i\theta})$ is a trigonometric polynomial of degree $N-1$, $P(\theta)/[\sin(b\theta/2)/\sin(\theta/2)]^{2N}$ is a trigonometric polynomial of the same degree $N-1$. ∎

Definition. This polynomial $P(\theta)/[\sin(b\theta/2)/\sin(\theta/2)]^{2N}$ is called the *secondary factor* of $P(\theta)$. This factor is denoted by $S(\theta)$.

The next step is to study the following class of integrals: $\int |\xi|^{\alpha} G(\xi)\, d\xi$ where α is a parameter between 0 and $2N$. We start with the value $\alpha = 2N$.

Lemma 7.3. *In the interpolation process of type (b, N), there is a constant C_1 such that, for any $n \geq 0$,*

$$\int_{(0,\pi b^n)} \xi^{2N} G(\xi)\, d\xi \leq C_1 b^{2Nn} \int_{(0,2\pi)} \prod_{0 \leq k \leq n-1} S(\theta b^k)\, d\theta.$$

Proof.

$$I_n = \int_{(0,\pi b^n)} \xi^{2N} G(\xi)\, d\xi = b^{n(2N+1)} \int_{(0,\pi)} \xi^{2N} G(\xi b^n)\, d\xi,$$

$$I_n = b^{n(2N+1)} \int_{(0,\pi)} \xi^{2N} \prod_{0 \leq k \leq n-1} [P(\xi b^k)/b] G(\xi)\, d\xi,$$

$$I_n = b^{2Nn} \int_{(0,\pi)} \xi^{2N} [\sin(b^n \xi/2)/\sin(\xi/2)]^N \left[\prod_{0 \leq k \leq n-1} S(\xi b^k) \right] G(\xi)\, d\xi.$$

Since the functions of ξ, $\xi^{2N}[\sin(b^n\xi/2)/\sin(\xi/2)]^{2N} G(\xi)$, are uniformly bounded on $[0, \pi]$, there is a bound C_1 such that

$$I_n \leq C_1 b^{2Nn} \int_{(0,2\pi)} \prod_{0 \leq k \leq n-1} S(\theta b^k)\, d\theta, \qquad n = 0, 1, 2, \ldots.$$ ∎

We need to study the behavior of another sequence of integrals. When $S(\theta)$ is a trigonometric polynomial and when b is an integer larger than 1, we set $J_n = \int_{(0,2\pi)} \prod_{0 \leq k \leq n-1} S(\theta b^k)\, d\theta$. There is a close connection between the sequence J_n, an infinite matrix B which we introduce immediately, and eigenvalues of a truncated part of B.

If $S(\theta) = \sum s_k e^{ik\theta}$, the entries of matrix B are given as $b_{j,k} = s_{j-bk}$. Let us describe some properties of this matrix.

Lemma 7.4. *If $S(\theta) = \sum s_k e^{ik\theta}$ is a trigonometric polynomial, if B is the matrix (s_{j-bk}), and if $\sum_k s_k^{(n)} e^{ik\theta}$ is the polynomial $\prod_{k=0,n-1} S(b^k\theta)$, then the (j, k)-entry of the matrix B^n is the number $s_{j-kb^n}^{(n)}$.*

Proof. The action of matrix B is better understood by using formal trigonometric series. If $V=(v_k)$, then we introduce the trigonometric series $Y(\theta)=\sum_k v_k e^{ik\theta}$. We do not care about the convergence of this series. If $(w_k)=W=BV$ and if $Z(\theta)=\sum_k w_k e^{ik\theta}$, then $Z(\theta)=S(\theta)Y(b\theta)$. If $(w_k^{(n)})=W_n=B^nV$ and if $Z_n(\theta)=\sum_k w_k^{(n)} e^{ik\theta}$, then $Z_n(\theta)=\prod_{0\le k\le n-1} S(b^k\theta)Y(b^n\theta)$. The kth column of matrix B^n can be found from the trigonometric polynomial $Z_n(\theta)$ if $Y(\theta)=e^{ik\theta}$. From that formula, it is possible to see that the (j, k)-entry of matrix B^n is the number $s_{j-kb^n}^{(n)}$ if the polynomial $\prod_{k=0,n-1} S(b^k\theta)$ is equal to $\sum_k s_k^{(n)} e^{ik\theta}$. ■

Definition. We say that a matrix $B=(b_{j,k})_{-\infty<k<\infty,-\infty<j<\infty}$ is *M-concentrated* if, for every j of $[-M, M]$ and for every k outside of $[-M, M]$, $b_{j,k}=0$.

Definition. The *M-truncation* of a matrix $B=(b_{j,k})_{-\infty<k<\infty,-\infty<j<\infty}$ is the matrix $(b_{j,k})_{-M\le k\le M,-M\le j\le M}$. This matrix is denoted by $[B]_M$.

Lemma 7.5. *If $S(\theta)=\sum s_k e^{ik\theta}$ is a trigonometric polynomial of degree N, if B is the matrix (s_{j-bk}), and if M is the integral part of $N/(b-1)$, then B is an M-concentrated matrix.*

Proof. If M is the integral part of $N/(b-1)$, let us consider two indices j and k of matrix B such that $|j|\le M$ and $|k|>M$. Then $|j-bk|\ge(M+1)b-M$ and $|j-bk|\ge M(b-1)+b>N$. So $s_{j-bk}=0$. B is an M-concentrated matrix. ■

Lemma 7.6. *If B and C are two M-concentrated matrices, BC is an M-concentrated matrix and $[BC]_M=[B]_M[C]_M$.*

We leave the proof to the reader.

Corollary 7.7. *If A is an M-concentrated matrix, then for any positive integral value of n, $[A^n]_M=[A]_M^n$.*

Theorem 7.8. *In an iterative process of type (b, N), we set M as the integral part of $(N-1)/(b-1)$ and J_n as $\int_{(0,2\pi)}\prod_{0\le k\le n-1} S(\theta b^k)\,d\theta$. If r is the spectral radius of $[B]_M$ and if s is the largest multiplicity of an eigenvalue of modulus equal to r of this matrix, then there is a constant C_2 such that, for any positive integral value n,*

$$|J_n|\le C_2 n^{s-1} r^n.$$

Proof. If the secondary factor of an iterative process of type (b, N) is $S(\theta)=P(\theta)/[\sin(b\theta/2)/\sin(\theta/2)]^{2N}=\sum s_k e^{ik\theta}$, matrix B is $(b_{j,k})=(s_{j-bk})$. The number J_n is $\int_{(0,2\pi)}\prod_{0\le k\le n-1} S(\theta b^k)\,d\theta$. J_n is the central element (entry corresponding to rank 0 and column 0) of matrix B^n. Since S is a trigonometric polynomial of degree $N-1$, matrix B is M-concentrated if M is equal to the integral part of $(N-1)/(b-1)$. According to Corollary 7.7, J_n is the central element of the truncated matrix $[B]_M^n$. If $\varphi(x)=\sum_{k\ge0} c_k x^k$ is the minimal polynomial of $[B]_M$, then $[B]_M^n\varphi([B]_M)=\sum_{k\ge0} c_k[B]_M^{n+k}=0$ for $n=0, 1, 2, \ldots$. From this matrix equation, it follows that the sequence J_n satisfies the linear recurrence $\sum_{k\geq 0} c_k J_{n+k} = 0$. As can be found in Gel'fond [5] or in other classical books, the sequence J_n is a finite linear combination of the following specific sequences. These sequences are $n^j\lambda^n$ where λ is a root of the minimal polynomial $\varphi(x)$ and j is a nonnegative integer which is smaller than the multiplicity of the root λ. If r is the largest modulus of a root of φ, and if s is the largest multiplicity of a root of φ of modulus equal to r, then each admissible sequence $n^j\lambda^n$ is bounded by a suitable multiple of $n^{s-1}r^n$. There is a constant C_2 such that, for any positive integral value n, $|J_n| \leq C_2 n^{s-1} r^n$. The number r is also the spectral radius of matrix $[B]_M$. ■

Corollary 7.9. *In the process of type* (b, N), *if* r *is the spectral radius of matrix* $[B]_M$, *there is a number* C *and an integer* s *such that, for any* $n \geq 1$,

$$\int_{(0,\pi b^n)} \xi^{2N} G(\xi)\, d\xi \leq C n^s r^n b^{2Nn}.$$

Theorem 7.10. *If* r *is the spectral radius of the associated matrix* $[B]_M$ *of an interpolation process of type* (b, N) *and if* α *is a positive real value such that* $rb^\alpha < 1$, *then* $\int |\xi|^\alpha G(\xi)\, d\xi < \infty$.

Proof.

$$K_n = \int_{(\pi b^{n-1}, \pi b^n)} \xi^\alpha G(\xi)\, d\xi,$$

$$K_n \leq (b^{n-1})^{\alpha - 2N} \int_{(\pi b^{n-1}, \pi b^n)} \xi^{2N} G(\xi)\, d\xi.$$

$\sum_{n\geq 0} K_n \leq b^{2N-\alpha} C \sum_{n\geq 0} (b^\alpha r)^n n^s < \infty$ since by hypothesis $b^\alpha r < 1$. So it follows that $\int |\xi|^\alpha G(\xi)\, d\xi = 2(\int_{(0,\pi)} \xi^\alpha G(\xi)\, d\xi + \sum_{n\geq 0} K_n) < \infty$. ■

Definition. The number E such that $b^E = 1/r$ is called the *critical exponent* of the process of interpolation. An application of Lemma 7.1 and Theorem 7.8 gives the following results.

Theorem 7.11. *If* m *is a positive integral value smaller than* E, *then the derivative of order* m *of the fundamental function is continuous. Moreover, if* m *is a nonnegative integral value and if* α *a is real number between* 0 *and* 1 *such that* $m + \alpha$ *is smaller than* E, *then* $F^{(m)}(t)$ *belongs to the Lipschitz class of order* α.

8. Finite Differences of the Fundamental Function

We found that the derivative of a suitable order of the fundamental function belongs to a Lipschitz class. We prove in this section that this result cannot be substantially improved. If $S(\theta) = \sum_k s_k e^{ik\theta}$ is the secondary factor of the characteristic polynomial $P(\theta)$, we have already defined a matrix $B = (s_{j-bk})$. We define another matrix from the fundamental function F: $A = (F(j/b - k))$. We begin with the study of eigenvalues of matrices A and B.

Lemma 8.1. *If r is an integer of $[0, 2N-1]$ and if V is the column vector $(k^r)_{-\infty<k<\infty}$, then $AV = b^{-r}V$.*

Proof. If r is an integer of $[0, 2N-1]$, it follows from Lemma 4.1 and Theorem 3.2 that $\sum_k F(j/b-k)k^r = (j/b)^r$. If V is the column vector $(k^r)_{-\infty<k<\infty}$, then $AV = b^{-r}V$. ∎

Theorem 8.2. *Any eigenvalue of matrix A other than the following powers of $(1/b)$: $1, 1/b, 1/b^2, \ldots, 1/b^{2N-1}$, is an eigenvalue of matrix B. Any eigenvalue of B is an eigenvalue of A.*

Proof. Let us assume that V is column-vector $\neq 0$, with complex components, and λ is a complex number such that $AV = \lambda V$. We also assume that λ is not one of the following integral powers of $1/b$: $1, 1/b, 1/b^2, \ldots, 1/b^{2N-1}$. If $V = (v_k)$, then we introduce the trigonometric series $Y(\theta) = \sum_k v_k e^{ik\theta}$. We do not care about the convergence of this series, it is a formal trigonometric series. Since $AV = \lambda V$, then $P(\theta)Y(b\theta) = \lambda Y(\theta)$ where $P(\theta)$ is the characteristic polynomial of the interpolation process. According to Theorem 7.2, $P(\theta) = [\sin(b\theta/2)/\sin(\theta/2)]^{2N} S(\theta)$.

If $Z(\theta) = \sin^{2N}(\theta/2)Y(\theta)$, then $S(\theta)Z(b\theta) = \lambda Z(\theta)$. If $Z(\theta) = \sum_k w_k e^{ik\theta}$ and if $W = (w_k)$, then $BW = \lambda W$. Let us check that $W \neq 0$. If $W = 0$, then we would get $Z(\theta) = 0 = \sin^{2N}(\theta/2)Y(\theta)$. The meaning of the relation $\sin^{2N}(\theta/2)Y(\theta) = 0$ is that all finite differences of order $2N$ of the sequence v_k are zero. If this is the case, then there would be a polynomial $p(t)$ of degree smaller than $2N$ such that $v_k = p(k)$, $k = 0, \pm 1, \pm 2, \ldots$. If $V_r = (k^r)$, then V would be a linear combination of $\{V_r\}_{0\le r\le 2N-1}$. As $AV_r = 1/b^r V_r$, $AV = \lambda V$ and λ is not one of the numbers $1/b_r (0 \le r \le 2N-1)$, then we would get a contradiction.

Let us now look at the second part of the theorem. If λ is an eigenvalue of B, we need to prove that λ is an eigenvalue of A. The case $\lambda = 1/b^r$ (where $r = 0, 1, \ldots, 2N-1$) is dealt with in Lemma 8.1. We may assume that λ is an eigenvalue that is different from $1/b^r$ $(r = 0, 1, \ldots, 2N-1)$. Then there is a vector $W = (w_k)$ such that $BW = \lambda W$. We set $Z(\theta) = \sum_k w_k e^{ik\theta}$ and the relation $S(\theta)Z(b\theta) = \lambda Z(\theta)$ is induced. If δ_{2N} is the operator of the central finite differences of order $2N$, then there is a sequence v_k such that $w_k = \delta_{2N} v_k$. If we set $Y(\theta) = \sum_k v_k e^{ik\theta}$, then $Z(\theta) = \sin^{2N}(\theta/2)Y(\theta)$. From this, it follows that $S(\theta)\sin^{2N}(b\theta/2)Y(b\theta) = \lambda \sin^{2N}(\theta/2)Y(\theta)$ and $\sin^{2N}(\theta/2)[P(\theta)Y(b\theta) - \lambda Y(\theta)] = 0$.

If $V = (v_k)$, then there is a polynomial $p(k)$ of degree smaller than $2N$ such that $AV - \lambda V = (p(k))$. It is easy to find a polynomial $q(k)$ of degree smaller than $2N$ such that $(A - \lambda I)(q(k)) = (p(k))$. The column vector $V - (q(k))$ is then an eigenvector for matrix A. ∎

Lemma 8.3. *If A is the matrix $(F(j/b-k))_{-\infty<k<\infty, -\infty<j<\infty}$, then matrix A^n is $(F(j/b^n - k))$.*

Proof. We proceed by induction on n. If $A^{n+1}=(c_{j,k})$, we use the relation $A^{n+1}=AA^n$.

$$c_{j,k}=\sum_r F(j/b-r)F(r/b^n-k)=\sum_s F((j-kb^{n+1})/b-s)F(s/b^n).$$

From equation (3.1), $c_{j,k}=F(j/b^{n+1}-k)$. ■

Lemma 8.4. *Any eigenvalue of B is in the unit-disk of the complex plane.*

Proof. If λ is an eigenvalue of matrix B, there is a rank vector $W=(w_k)_{-M\le k\le M}$ which is an eigenvector for the truncated matrix $[A]_M$: $W[A]_M=\lambda W$. Since $W[A]_M^n=\lambda^n W$, it follows that

$$\sum_j w_j F(j/b^n-k)=\lambda^n w_k \qquad (\text{for } k=-M,-M+1,\ldots,M-1,M).$$

Let us assume that $|\lambda|$ is larger than 1. The vector W is biorthogonal to the eigenvector $(1)_{-M\le k\le M}$ corresponding to the eigenvalue 1: $\sum_k w_k=0$. If $s_k=\sum_{j\le k} w_j$, then $\sum_j w_j F(j/b^n-k)=-\sum_j s_j[F((j+1)/b^n-k)-F(j/b^n-k)]$.

If k is an index for which $w_k\ne 0$, it is possible to find a positive number ε and a sequence j_n of integers such that $F((j_n+1)/b^n-k)-F(j_n/b^n-k)$ is larger than $\varepsilon|\lambda|^n$ in absolute value. The number ε can be chosen as $|w_k/[(2M+1)B]|$ where B is the maximum of numbers $|s_j|$. Since F is continuous, then $|\lambda|<1$. This is a contradiction. ■

Theorem 8.5. *If E is the critical exponent of the interpolation process, we set m as the integral part of E and $\alpha=E-m$. If E is not integral, $F^{(m)}(t)$ cannot belong to the Lipschitz class* Lip *s when s is larger than α. If E is integral, $F^{(m)}$ cannot be a continuous function.*

Proof. Let us first discuss the case where E is not integral. If λ is the maximal eigenvalue of matrix B, there is a rank vector $W=(w_k)_{-M\le k\le M}$ which is an eigenvector for the truncated matrix $[A]_M$: $W[A]_M=\lambda W$. Since $W[A]_M^n=\lambda^n W$, it follows that

$$\sum_j w_j F(j/b^n-k)=\lambda^n w_k \qquad (\text{for } k=-M,-M+1,\ldots,M-1,M).$$

The vector W is biorthogonal to each column eigenvector corresponding to the eigenvalue $1/b^r$, $r=0,1,\ldots,m$. W is then biorthogonal to any column vector $V=(p(j))_{-M\le j\le M}$ where p is a polynomial of degree smaller or equal to m. From this fact, it follows that W is a linear combination of operators of finite difference of order $m+1$ (any operator of finite difference can be viewed as a rank vector). Let us pick an index k for which $w_k\ne 0$. It is possible to find a positive number ε and a sequence D_n of operators of finite difference of order $m+1$ such that when D_n is applied to the sequence $V_n=(F(j/b^n-k))_{-M\le j\le M}$, a number D_nV_n larger than $\varepsilon\lambda^n$ in absolute value is then produced. The number ε can be chosen as $|w_k/[(2M+1)B]|$ if B is a bound of the absolute values of the coefficients used in the representation of W as a linear combination of operators of finite difference of order $m+1$.

For each value of n, there is a couple of real numbers (u_n, t_n) such that $D_n V_n = [F^{(m)}(t_n) - F^{(m)}(u_n)]b^{-nm}$ and $|u_n - t_n| \leq 2M/b^n$. With these inequalities, there is no way for the function $F^{(m)}$ to belong to Lipschitz class Lip s if s is larger than α. One part of the proof is complete.

Let us discuss the case where E is integral. The maximal eigenvalue λ of matrix B is $1/b^m$.

(a) If there is a rank vector $W = (w_k)_{-M \leq k \leq M}$ which is at the same time an eigenvector for the truncated matrix $[A]_M$: $W[A]_M = \lambda W$ and is biorthogonal to the column eigenvector $(k^m)_{-M \leq k \leq M}$, then the same arguments as before show that $F^{(m)}$ cannot be continuous. A sequence of couples of real numbers (u_n, t_n) can be found such that $F^{(m)}(t_n) - F^{(m)}(u_n)$ does not converge to 0 while $|u_n - t_n| \leq 2M/b^n$.

(b) Otherwise, there is a row eigenvector W and a row vector W' such that $W[A]_M = \lambda W$, $W'[A]_M = \lambda W' + W$, and W' is biorthogonal to the column eigenvector $(k^m)_{-M \leq k \leq M}$. If $W = (w_j)$ and $W' = (w_j')$, then $W[A]_M^n = \lambda^n W$ and $W'[A]_M^n = \lambda^n W' + n\lambda^{n-1} W$.

$$\sum_k w_j' F(j/b^n - k) = \lambda^n w_k' + n\lambda^{n-1} w_k \qquad (\text{for } k = -M, -M+1, \ldots, M-1, M).$$

Similar arguments show that $F^{(m)}$ cannot be continuous. There is a positive number ε and a sequence of couples of real numbers (u_n, t_n) can be found such that $|F^{(m)}(t_n) - F^{(m)}(u_n)| > \varepsilon n$ while $|u_n - t_n| \leq 2M/b^n$. ■

9. Examples of Secondary Factors

To illustrate previous theorems, we give examples of secondary factors. We bring back the polynomial

$$\Pi(w) = \sum_{k=-bN+1, bN-1} F(k/b) w^{k+2Nb-1}.$$

We have shown that $\Pi(w)$ is divisible by $[(w^b - 1)/(w-1)]^{2N}$, the quotient is a polynomial $\Sigma(w)$ of degree $2N-2$. We compute the first two coefficients σ_0 and σ_1 of $\Sigma(w)$.

Lemma 9.1. *In the interpolation process of type* (b, N),

$$\sigma_0 = (1/b)^{2N-1}\left[\prod_{k=1,N-1}(1 - k^2 b^2)\right]/(2N-1)!,$$

$$\sigma_1 + 2N\sigma_0 = 2(1/b)^{2N-1}\left[\prod_{k=1,N-1}(4 - k^2 b^2)\right]/(2N-1)!.$$

Proof. We used the fact that the first two coefficients of $\Pi(w)$ are $F(-bN + 1/b)$ and $F(-bN + 2/b)$. According to formula (4.2),

$$F(-bN + r/b) = \left[\prod_{k=-N+1, N-1}(r/b - k)\right]\bigg/(2N-1)!.$$

The first two coefficients of $[(w^b-1)/(w-1)]^{2N}$ are 1 and $2N$. Since $(\sigma_0+\sigma_1 w+\cdots)(1+2Nw+\cdots)=\Pi(w)$, a simple computation concludes the proof. ■

Lemma 9.2. *For an interpolation process of type* $(b, 2)$, *the secondary factor is* $S(y)=[(b^2+2)-(b^2-1)\cos y]/(3b^3)$.

Lemma 9.3. *For an interpolation process of type* $(b, 3)$, $S(y)$ *is* $[3(4b^4+5b^2+11)-2(b^2-1)(8b^2+13)\cos y+(b^2-1)(4b^2-1)\cos(2y)]/(60b^5)$.

Last two lemmas can be proved through Lemma 9.1 and with a little algebra.

We give the first secondary factors S (when $N=2$) for $b=2, 3, 4$. If $b=2$, $S(y)=1/4-\cos y/8$. If $b=3$, $S(y)=(11-8\cos y)/81$. If $b=4$, $S(y)=(6-5\cos y)/64$.

Let us look at eigenvalues of matrix $[B]_M$ as previously described. If $b=2$, then $[B]_M$ is a matrix of order 3 whose eigenvalues are 1/4 and −1/16. The critical exponent is in this case $E=2$. This means that the fundamental function is differentiable and F' belongs to Lipschitz class Lip s for any $s\in(0,1)$.

If $b>2$, then the matrix $[B]_M$ is of order 1. The unique eigenvalue λ is the constant coefficient in the Fourier expansion of $S(y)$, $\lambda=(b^2+2)/(3b^3)$. The critical exponent E is the solution to equation $b^E=3b^3/(b^2+2)$. E is between 1 and 2. The function F always has a continuous derivative. For example, if $b=3$, then $E=1.817$ and if $b=4$, then $E=1.707$.

We give the first secondary factors S (when $N=3$) for $b=2, 3, 4$:

$$S(y)=[19-18\cos y+3\cos(2y)]/128 \qquad \text{if} \quad b=2.$$

$$S(y)=[57-68\cos y+14\cos(2y)]/729 \qquad \text{if} \quad b=3.$$

$$S(y)=[223-282\cos y+63\cos(2y)]/4096 \qquad \text{if} \quad b=4.$$

Let us look at eigenvalues of matrix $[B]_M$ as previously described. If $b=2$, then $[B]_M$ is a matrix of order 5 whose eigenvalues are 3/256, −9/128, 9/64, and −1/16. The largest eigenvalue of matrix $[B]_M$ is 9/64. The critical exponent in this case is $E=\log_2(64/9)=2.830$. This means that the fundamental function is twice differentiable and F'' belongs to Lipschitz class Lip s for any $s\in(0, 0.830)$.

If $b=3$, then $[B]_M$ is a matrix of order 3 whose eigenvalues are 7/729 and 57/729. The critical exponent is then 2.319. F'' belongs to Lipschitz class Lip s for any $s\in(0, 0.319)$. If $b>3$, than $[B]_M$ is of order 1. The unique eigenvalue λ is the constant coefficient in the Fourier expansion of $S(y)$: $(4b^4+5b^2+11)/20b^5)$. The critical exponent E is the root of the equation $b^E=1/\lambda$. If $b=4$, then $E=2.099$, F'' belongs to Lip s for any $s\in(0, 0.099)$. If $b\geq 5$, then $1<E<2$.

10. The Derivatives of the Fundamental Function

If $F(t)$ is the fundamental function and if r is a positive integer smaller than the critical exponent E, then we will indicate how to compute the rth derivative of F, $F^{(r)}(t)$, when t is a b-adic rational number. As we show immediately, the main point is the computation of $F^{(r)}(t)$ at integral values of t. This is a consequence of formula (3.1). If we differentiate both sides of this formula r times with respect to t, then we get

$$F^{(r)}(k/b^m)/b^{rm} = \sum_n F(n/b^m)F^{(r)}(k-n). \tag{10.1}$$

We consider the case of an interpolation process of type (b, N). $F(t)$ is an even function and $F'(t)$ is an odd function. In particular, $F'(0)=0$. $F^{(r)}$ is an even function if r is even and $F^{(r)}$ is odd when r is odd. We find the relationship between $F^{(r)}(t)$ at integral values of t and row eigenvectors of the matrix $A=(F(j/b-k))$. We compute $F'(n)$ for every integer n, for every base b, for $N=2$ and 3. We compute $F''(n)$ for every integer n, for every base b, for $N=3$.

Theorem 10.1 *If r is a nonnegative integer smaller than the critical exponent of an interpolation process of type (b, N), then the row vector $(F^{(r)}(n))_{-\infty<n<\infty}$ is an eigenvector of the matrix $A=(a_{j,k})=(F(j/b-k))$ for the eigenvalue $1/b^r$. Moreover, $\sum_n n^r F^{(r)}(n)=(-1)^r r!$.*

Proof. If $V=(F^{(r)}(j))_{-\infty<j<\infty}$ and $(w_j)=VA$, then

$$w_k = \sum_j F^{(r)}(j)F(j/b-k) = \sum_n F^{(r)}(n+kb)F(n/b).$$

If, in the last member, n is changed to $-n$ and if the symmetry of F is used, then $w_k = \sum_n F^{(r)}(kb-n)F(n/b)$. From formula (10.1),

$$w_k = F^{(r)}(k/b)/b^r = v_k/b^r.$$

W is an eigenvector and $1/b^r$ is the eigenvalue. In order to prove the last part of the theorem, we use Lemma 4.1 with the polynomial $P(t)=t^r$ and Theorem 3.2: $t^r = \sum_n n^r F(t-n)$. If this identity is differentiated r times with respect to t, then

$$r! = \sum_n n^r F^{(r)}(t-n) = (-1)^r \sum_n n^r F^{(r)}(t+n).$$

If t is set to zero, then $\sum_n n^r F^{(r)}(n)=(-1)^r r!$. ■

Corollary 10.2. *If r is a nonnegative integer smaller than the critical exponent of an interpolation process of type (b, N), if M is the largest integer smaller than $N+(n-1)/(b-1)$, if $V=(v_j)_{-M\le j\le M}$ is a row eigenvector of the truncated matrix $[A]_M=[(F(j/b-k))]_M$ for the eigenvalue $1/b^r$, and if $\sum_n n^r v_n=(-1)^r r!$, then $V=(F^{(r)}(n))$.*

This follows from the fact that $1/b^r$ is a simple eigenvalue of matrix $[A]_M$.

Theorem 10.3. *In the interpolation process of type $(b, 2)$, $F'(1)=-2/3$ and $F'(2)=1/12$ whatever the base b.*

Proof. From Lemma 3.1, $F(t)$ vanishes on $[2+1/(b-1), \infty)$. $F(t)=0$ on $[3, \infty)$ and $F'(3)=0$. In order to compute $F'(2)$, Theorem 3.2 and Lemma 4.1 are invoked; if $p(t)=t(t^2-1)$, the following identity holds when $|t|<1$:

$$-p(t)=p(2)[F(t+2)-F(t-2)]+p(3)[F(t+3)-F(t-3)].$$

If we differentiate both sides with respect to t and if we set t to 0, $F'(2)=-p'(0)/(2p(2))=1/12$.

Computation of $F'(1)$ is performed using a similar argument. If $q(t)=t(t^2-4)$ and if $|t|<1$, then $-q(t)=q(1)[F(t+1)-F(t-1)]+q(3)[F(t+3)-F(t-3)]$. So $F'(1)=-q'(0)/(2q(1))=-2/3$. ■

Theorem 10.4. *In the interpolation process of type* $(b, 3)$ *with a base* $b \geq 3$, $F'(1)=-3/4$, $F'(2)=3/20$, and $F'(3)=-1/60$.

Proof. We argue as in the previous theorem. From Lemma 3.1, $F(t)$ vanishes on $[3+2/(b-1), \infty)$. Since $b \geq 3$, then $F(t)=0$ on $[4, \infty)$. $F'(4)=0$. In order to compute $F'(3)$, Theorem 3.2 and Lemma 4.1 are again invoked; if $p(t)=t(t^2-1)(t^2-4)$, the following identity holds when $|t|<1$:

$$-p(t)=p(2)[F(t+2)-F(t-2)]+p(3)[F(t+3)-F(t-3)]$$
$$+p(4)[F(t+4)-F(t-4)].$$

Hence $F'(3)=-p'(0)/(2p(3))=-1/60$.

Computation of $F'(2)$ is performed using the polynomial $q(t)=t(t^2-1)(t^2-9)$. $F'(2)=-q'(0)/(2q(2))=3/20$. Computation of $F'(1)$ is done by using the polynomial $r(t)=t(t^2-4)(t^2-9)$. $F'(1)=-r'(0)/(2r(1))=-3/4$. ■

For computation of the derivatives of F at integral points for six nodes with base 2, there is a small complication because $F'(4) \neq 0$. Using the same polynomials p, q, and r as in the previous proof, we get three equations relating $F'(1)$, $F'(2)$, $F'(3)$, and $F'(4)$:

$$F'(1)+r(4)F'(4)/r(1)=-r'(0)/(2r(1)),$$
$$F'(2)+q(4)F'(4)/q(2)=-q'(0)/(2q(2)),$$

and

$$F'(3)+p(4)F'(4)/p(3)=-p'(0)/(2p(3)).$$

Another equation can be found by using the fact that $F(t+4)=-3F(2t+3)/256$ when t is positive. So $F'(4)=-3F'(3)/128$. The solution of these linear equations is the content of next theorem.

Theorem 10.5. *In the interpolation process of type* $(2, 3)$, $F'(1)=-272/365$, $F'(2)=53/365$, $F'(3)=-16/1095$, *and* $F'(4)=-1/2920$.

Interpolation processes of type (b, N) for which F is twice differentiable occur in the following cases: $N=3$ and $b=2$, 3, or 4. After some computation, the following results about $F''(n)$ can be proved.

Theorem 10.6. *In the interpolation process of type* $(2, 3)$ (*the base is* 2), $F''(0) = -295/56$, $F''(1) = 356/105$, $F''(2) = -92/105$, $F''(3) = 4/35$, *and* $F''(4) = 3/560$.

Theorem 10.7. *In the interpolation process of type* $(b, 3)$ *with a base* $b = 3$ *or* 4,

$$F''(0) = (25/18)(4b^4 + 36b^3 - 17b^2 - 23)/(4b^4 - 20b^3 + 5b^2 + 11),$$

$F''(1) = -13/24 - 3F''(0)/4$, $\quad F''(2) = 2/3 + 3F''(0)/10$, $\quad F''(3) = -1/8 - F''(0)/20$ *and, for* $n \geq 4$, $F''(n) = 0$.

11. Error Bounds in Interpolation

Let $f(t)$ be a real function defined on the real axis. We use the following sequence: $y(n) = f(n)$ for relative integers n. $y(t)$ is the interpolation of the sequence $y(n)$ according to scheme (4.1) for a given base b and with $2N$ nodes. The error in the interpolation is the function $e(t) = f(t) - y(t)$. We would like to get simple bounds for $e(t)$.

Theorem 11.1. *Let* $y(t)$ *be the interpolating function for a function* f *according to scheme* (4.1) *with a base* b *and an even number* $2N$ *of nodes. If* $t \in [0, 1]$ *and if* $p(t)$ *is the polynomial of degree* $2N - 1$ *such that* $p(n) = f(n)$, $n = -N+1, -N+2, \ldots, N-1$, *and* N, *then*

$$|f(t) - y(t)| \leq |f(t) - p(t)| + \sum_{k>0} (|f(N+k) - p(N+k)|$$
$$+ |f(-N+1-k) - p(-N+1-k)|)\varepsilon_k.$$

The number ε_k *is the maximal value of* $|F(t)|$ *on* $[k-1, k]$.

Proof. We use the triangular inequality $|f(t) - y(t)| = |f(t) - p(t) + p(t) - y(t)| \leq |f(t) - p(t)| + |p(t) - y(t)|$. If $t \in [0, 1]$, then Theorem 3.2 tells us that $p(t) - y(t) = \sum_k [p(k) - f(k)]F(t-k)$. Since $p(k) = f(k)$ for $k = -N+1, -n+2, \ldots, N-1, N$ and since F is an even function, then

$$p(t) - y(t) = \sum_{k>N} \{[p(k) - f(k)]F(k-t) + [p(-k+1) - f(-k+1)]F(k-1+t)\},$$

$$|p(t) - y(t)| \leq \sum_{k>N} (|p(k) - f(k)| + |p(-k+1) - f(-k+1)|)\varepsilon_k.$$

The required inequality is then proved. ■

A similar argument can be used to obtain bounds about the rth derivative of $y(t) - p(t)$. If $\varepsilon_k^{(r)}$ is the maximal value of $|F^{(r)}(t)|$ on $[k-1, k]$, then

$$|f^{(r)}(t) - y^{(r)}(t)| \leq |p^{(r)}(t) - y^{(r)}(t)|$$
$$+ \sum_{k>N} (|p(k) - f(k)| + |p(-k+1) - f(-k+1)|)\varepsilon_k^{(r)}.$$

We give a table of numbers $\varepsilon_k^{(r)}$ for $b = 2$, 3, and 4 and for $N = 2$ and 3. These numbers were found after numerical experiments.

Case $N=2$

b	ε_3	$\varepsilon_3^{(1)}$
2	0.00459	0.08333
3	0.00316	0.08333
4	0.00251	0.08333

Case $N=3$

b	ε_4	ε_5	$\varepsilon_4^{(1)}$	$\varepsilon_5^{(1)}$	$\varepsilon_4^{(2)}$	$\varepsilon_5^{(2)}$
2	0.00122	0.00001	0.01461	0.00034	0.11429	0.00536
3	0.00083	0	0.16667	0	0.36111	0
4	0.00063	0	0.16667	0	1.15152	0

Remark. When $N=2$, it seems that the exact value of $\varepsilon_3^{(1)}$ is $1/12$ which is the value of $F'(2)$. When $N=3$, the exact values of $\varepsilon_4^{(1)}$, $\varepsilon_5^{(1)}$, $\varepsilon_4^{(2)}$, and $\varepsilon_5^{(2)}$ are probably $|F'(3)|$, $|F'(4)|$, $|F''(3)|$, and $|F''(4)|$, respectively.

12. Conclusion

The main purpose of this paper was to study a process of interpolation which could be useful for numerical analysis. This paper is a sequel to three other works, [4], [2], and [3]. In [4] the second author introduced the interpolation process of type (2, 2), the base is 2, the number of nodes is four. The fundamental function F of that case was then studied; in [2] we indicated that some properties of F could be derived more easily by using Fourier transform. In [2] and in [3] we defined what we call the model of iterative interpolation with a base b. As has been noted, this model embodies all von Koch-Mandelbrot curves as described in Mandelbrot's book [6]. We say that an iterative interpolation process is in L^2 if the Fourier transform of the process is in L^2. In [3] we found all von Koch-Mandelbrot curves which are in L^2.

If there is a choice to be done amongst a base b, it seems better to take the base b equal to 2. We can expect that the interpolation which will be produced will be smoother than with another base. If the complexity of the computations is increased when a large number of moving nodes is used, however, the accuracy of the interpolation will probably be improved with "analytic" data. It should be interesting to know the behavior of the interpolation process of type $(2, N)$ for other values of N such as $N=4$, 5, and 6.

We now present some open questions. Is it true that any interpolation process of type (b, N), $N \geq 2$, gives rise to a continuously differentiable fundamental function? If E is the critical exponent of an interpolation process of type (b, N) and if m is the integral part of E, is it possible to find the Hausdorff dimension of the graph of $F^{(m)}$? The last question is the following. How can we handle the study of the general iterative interpolation process when it is not related to polynomial interpolation? In (2.1) we may take parameters p_k as reals or even as complex numbers. The main question is to find conditions on the parameters p_k such that the interpolating function $y(t)$ defined on b-adic numbers has a continuous extension to the real axis. It does not seem easy to complete this task satisfactorily. We conjecture that the fundamental function has a continuous extension if the value 1 is a simple eigenvalue of matrix $A = (F(j/b - k))$ and if the modulus of every other eigenvalue is smaller than 1.

Acknowledgments. Both authors were supported by FCAR (CRP-2093, Ministère de l'Éducation du Québec). The second author was supported by NSERC (Operating Grant No. A8860, Federal Government of Canada).

References

1. S. Bochner, K. Chandrasekharan (1949): Fourier Transforms. Princeton: Princeton University Press.
2. G. Deslauriers, S. Dubuc (1987): *Interpolation dyadique.* In: Fractals. Dimensions non entières et applications. Paris: Masson, pp. 44–55.
3. G. Deslauriers, S. Dubuc (1987): *Transformées de Fourier de courbes irrégulières.* Ann. Sci. Math. Québec, **11**:25–44.
4. S. Dubuc (1986): *Interpolation through an iterative scheme.* J. Math. Anal. Appl., **114**:185–204.
5. A. O. Gel'fond (1971): Calculus of Finite Differences. Delhi: Hindustan.
6. B. B. Mandelbrot (1982): The Fractal Geometry of Nature. San Francisco: Freeman.

G. Deslauriers
Département de mathématiques appliquées
École Polytechnique
C.P. 6079, Succ. A
Montréal
Québec H3C 3A7
Canada

S. Dubuc
Département de mathématiques et de statistique
Université de Montréal
C.P. 6128, Succ. A
Montréal
Québec H3C 3J7
Canada

Constr. Approx. (1989) 5: 69–87

CONSTRUCTIVE APPROXIMATION

Approximation of Measures by Markov Processes and Homogeneous Affine Iterated Function Systems

John H. Elton and Zheng Yan

Abstract. It is shown that under certain conditions, attractive invariant measures for iterated function systems (indeed for Markov processes on locally compact spaces) depend continuously on parameters of the system.

We discuss a special class of iterated function systems, the homogeneous affine ones, for which an inverse problem is easily solved in principle by Fourier transform methods. We show that a probability measure μ on $\mathbf{R}^n$ can be approximated by invariant measures for finite iterated function systems of this class if $\hat{\mu}(t)/\hat{\mu}(a^T t)$ is positive definite for some nonzero contractive linear map $a: \mathbf{R}^n \to \mathbf{R}^n$. Moments and cumulants are also discussed.

Introduction

There has been much interest lately in producing images, which may be thought of as probability measures on $\mathbf{R}^2$, by a random process. The idea is to recognize the image as the invariant measure for some Markov process which converges in distribution to this measure. This is valuable for image compression if the transition probability of the process is fairly simple, even though the image itself is complicated [BS].

This "inverse" problem, finding a reasonably simple process which reproduces (approximately) a given image, is a difficult and interesting one. Our main object here is to discuss a special class of measures for which this inverse problem can be solved using Fourier transform methods. These are the probability measures which can be approximated by finite homogeneous affine iterated function systems (see Section 3 for the definition).

We found it useful to give a general discussion of the continuous dependence of invariant measures for Markov processes on their transition probabilities under certain conditions (Section 1), and a rather general definition and discussion of iterated function systems (Section 2); this is done on a separable, locally compact metric space. This continuous dependence is exploited to prove the approximation result mentioned above.

Date received: May 20, 1987. Communicated by Michael F. Barnsley.
AMS classification: 60J05, 60J15, 60F05, 42A38, 42A82, 58F11, 41A99.
Key words and phrases: Invariant measures, Markov processes, Iterated function systems, Approximation of measures, Fourier transform, Positive definite.

1. Markov Processes and Invariant Measures

In this section we discuss discrete-time Markov processes with attractive invariant measures in general on a separable, locally compact metric space X. Our object is to prove that under certain conditions, attractive invariant measures depend continuously on the transition probabilities which generate them. This is applied later to show that attractive invariant measures for iterated function systems depend continuously on parameters, under the right conditions (which are easily seen to be satisfied for the homogeneous affine case we discuss in detail).

A. Let $\mathcal{P}(X)$ be the Borel probability measures on X, with the usual topology of weak convergence, or convergence in distribution; that is, for $\nu_n, \nu \in \mathcal{P}(X)$, $\nu_n \to \nu$ iff $\int f\,d\nu_n \to \int f\,d\nu$ for all $f \in C(X)$, the *bounded* real-valued continuous functions on X. (This may also be called w^*-convergence, since $\mathcal{P}(X) \subset C(X)^*$.)

A *transition probability* is a function $p: X \to \mathcal{P}(X)$ such that $p(\cdot, B)$ is a Borel-measurable function on X for each Borel set B in X (we write $p(x, B)$ as is commonly done, rather than $p(x)(B)$). We also always assume that p is a continuous function from $X \to \mathcal{P}(X)$. $p(x, B)$ represents the probability of transfer from x into B in one step of the process.

Define the usual associated operator on bounded real-valued Borel functions on X,

$$Tf(x) = \int f(y)p(x, dy).$$

Our assumption that p is continuous is equivalent to assuming T takes $C(X)$ into $C(X)$, as is easily seen. Such processes are called *Feller* processes. The probabilistic interpretation of T is

$$Tf(x) = \mathbf{E}f(\mathbf{Z}_1^x),$$

where $\{Z_n^x\}_{n=0}^{\infty}$ is the Markov stochastic process with initial distribution concentrated at x (i.e., $Z_0^x \equiv x$) and transition probability p. Similarly,

$$T^n f(x) = \mathbf{E}f(Z_n^x).$$

This is sometimes a useful interpretation.

A measure $\mu \in \mathcal{P}(X)$ is called *invariant* if $\int Tf\,d\mu = \int f\,d\mu$ for all $f \in C(X)$, or what is the same thing, $T^*\mu = \mu$, i.e., μ is a fixed point for the adjoint operator. In probabilistic terms, this means that the process started with initial distribution μ will continue to have distribution μ.

μ will be called *attractive* if $T^{*n}\nu \to \mu$ for all $\nu \in \mathcal{P}(X)$. This is equivalent to $\int T^n f\,d\nu \to \int f\,d\mu$ for all $f \in C(X)$, all $\nu \in \mathcal{P}(X)$. In probabilistic language, μ is attractive iff the Markov process converges in distribution to μ for all initial distributions. We use this word rather than the word *ergodic*, since ergodic is used in so many different ways, not equivalent to attractiveness in general.

It is easy to see that an attractive probability for a Feller process must be invariant (see [BE]), and this is the only kind of process we discuss.

A family of probability measures Λ on a metric space is called *tight* if for all $\varepsilon > 0$, there exists a compact set K such that $\lambda(\tilde{K}) < \varepsilon$ for all $\lambda \in \Lambda$.

Our process $\{Z_n^x\}$ is called tight if the corresponding distributions $\lambda_n(B) = \mathcal{P}(Z_n^x \in B)$ are a tight collection. Since $T^n f(x) = \mathbf{E}f(Z_n^x)$, it is easy to see (by Urysohn's lemma) that this has the following equivalent formulation in terms of the operator T:

The process started at x is *tight* iff for all $\varepsilon > 0$, there is a compact set $K \subset X$ such that

$$|T^n g(x)| \leq \|g\|_K + \varepsilon \|g\|$$

for all n, all $g \in C(X)$ (where $\|g\|_K = \sup_{x \in K} |g(x)|$ and $\|g\| = \|g\|_X$).

B. We now discuss what existence of an attractive invariant measure means in terms of T. Let $C_c(X)$ be the continuous real-valued functions on X with compact support.

Proposition 1. *There is an attractive probability measure for the process iff*

(i) *for all $f \in C_c(X)$, $T^n f$ converges pointwise to a constant (depending on f), and*

(ii) *for some x_0, the process started at x_0 is tight (it then follows that this is true for every x).*

Proof. Assume $\mu \in \mathcal{P}(X)$ such that $T^{*n}\nu \to \mu$ for all $\nu \in \mathcal{P}(X)$. Taking ν to be the Dirac measure δ_x, and $f \in C_c(X)$,

$$\int T^n f \, d\delta_x = T^n f(x) \to \int f \, d\mu,$$

which is independent of x. It follows by the easy half of Prohorov's theorem that the process starting at any x is tight since it converges in distribution.

Conversely, assume $T^n f(x) \to L(f)$ (independent of x) for every $f \in C_c(X)$. This is clearly a bounded positive linear functional, so we may write $T^n f(x) \to \int f \, d\mu$ for some positive Borel measure μ, by Riesz's representation theorem. By tightness of the process started at x_0, it follows by a standard argument that μ is a probability measure. Thus for any $\nu \in \mathcal{P}(X)$, $\int T^n f \, d\nu \to \int (\int f \, d\mu) d\nu = \int f \, d\mu$, by Lebesgue's theorem, and this is easily extended to $C(X)$ since μ is a probability measure, so $T^{*n}\nu \to \mu$. ∎

The next proposition, which gives a useful way of deciding when a Markov process on X has an attractive measure, is motivated by the method of proof in [BDEG] showing existence of an attractive measure under certain average contractivity conditions, extended to infinitely many maps in [BA]. Our proof here, simpler than the corresponding part of the proof in [BDEG], was motivated by the proof in [BA], using Ascoli's theorem.

Proposition 2. *Assume*

(i) *for any* $f \in C_c(X)$, $\{T^n f, n = 0, 1, 2, \ldots\}$ *is equicontinuous*;
(ii) $\lim_{n\to\infty} |T^n f(x) - T^n f(y)| = 0$ *for all* $f \in C_c(X)$, $x, y \in X$; *and*
(iii) *the process started at some* x_0 *is tight.*

Then, for all $f \in C_c(X)$, $T^n f \to$ *constant (depending on* f*) uniformly on compact sets, so there is an attractive probability measure, by Proposition* 1.

Proof. Let $f \in C_c(X)$. By Ascoli's theorem, for some subsequence, $T^{n_j} f$ converges uniformly on compacts, to what must be a constant function, say c, by (ii). Let $\varepsilon > 0$; choose a compact K as in the definition of tightness and take j so large that $\|T^{n_j} f - c\|_K < \varepsilon$; then $|(T^n f - c)(x_0)| = |T^{n-n_j}(T^{n_j} f - c)(x_0)| \leq \varepsilon + \varepsilon(\|f\| + |c|)$ for $n \geq n_j$. Thus $T^n f(x_0) \to c$, hence $T^n f(x) \to c$ for all x, by (ii). ■

C. Next we discuss how the invariant measure μ varies when p is varied. We assume that p and p_k, $k = 1, 2, \ldots$, are transition probabilities giving attractive probability measures μ and μ_k, $k = 1, 2, \ldots$, and that $p_k(x, \cdot) \to p(x, \cdot)$ for all x. In addition, we assume that $\{p_k, k = 1, 2, \ldots\}$ is *equicontinuous*; that is,

$$\lim_{x\to x_0} \sup_k \left| \int f(y) p_k(x, dy) - \int f(y) p_k(x_0, dy) \right| = 0$$

for each $f \in C(X)$, or, equivalently, $\{T_k f, k = 1, 2, \ldots\}$ is equicontinuous for each $f \in C(X)$, where T_k is the operator associated with p_k. An equivalent condition, perhaps easier to check, is that $\{T_k f\}$ is equicontinuous for each $f \in C_c(X)$ *and* $\{p_k(x, \cdot), k = 1, 2, \ldots\}$ is tight for all x.

Note that $\mathcal{P}(X)$ is metrizable since X is separable, so by a general version of Ascoli's theorem [D, pp. 267 and 276], equicontinuity of $\{p_k\}$ is equivalent to saying $\{p_k\}$ is a relatively compact subset of $\mathcal{P}(X)^X$ in the compact-open topology, i.e., the topology of uniform convergence on compact sets in X. Thus our assumption on $\{p_k\}$ is that $p_k \to p$ in the C-O topology on $\mathcal{P}(X)^X$.

Lemma 1. *Under the assumptions above, for all n, for all* $f \in C(X)$, $T_k^n f \to^{k\to\infty} T^n f$ *uniformly on compact sets.*

Proof. First we prove

Claim. For any compact set $C \subset X$, for all $\varepsilon > 0$, there exists compact $K \subset X$ such that $\|T_k g\|_C \leq \|g\|_K + \varepsilon\|g\|$ for all k, all $f \in C(X)$.

Proof. *Let* $x \in C$. Since $\{p_k\}$ is relatively compact in $\mathcal{P}(X)^X$, $\{p_k(x, \cdot)\}$ is relatively compact in $\mathcal{P}(x)$, so is *tight*, by Prohorov's theorem. So there exists compact $K_x \subset X$ such that $p_k(x, \tilde{K}_x) < \varepsilon/2$ for all k. By equicontinuity, an easy argument using Urysohn's lemma, enlarging K_x, if necessary, yields $\delta_x > 0$ such that $p_k(y, \tilde{K}_x) < \varepsilon$ for all $y \in B_{\delta_x}(x)$, all k (where $B_\delta(x)$ is the ball of radius r centered at x). Since C is compact, it is covered by finitely many balls $B_{\delta_{x_i}}(x_i)$, $i = 1, \ldots, n$; letting $K = \bigcup_{i=1}^n K_{x_i}$, we have $p_k(y, \tilde{K}) < \varepsilon$ for all k, all $y \in C$. From the definition of T_k, this obviously implies the claim.

Proof of Lemma. Since $p_k \to p$, we have by equicontinuity $T_k f \to Tf$ uniformly on compact sets for each $f \in C(X)$.

Make the induction hypothesis that $T_k^{n-1} f \to T^{n-1} f$ uniformly on compact sets, for all $f \in C(X)$. Now let C be a compact set in X, and let $\varepsilon > 0$. Then

$$T_k^n f - T^n f = T_k(T_k^{n-1} f - T^{n-1} f) + (T_k - T)(T^{n-1} f).$$

Choose k_0 so that $k \geq k_0 \Rightarrow \|(T_k - T)(T^{n-1} f)\|_C < \varepsilon$ (we can do this since $T_k g \to Tg$ uniformly on compacts, where $g = T^{n-1} f$). Let k be the compact set in the claim; by induction hypothesis, there exists $k_1 \geq k_0$ such that $k \geq k_1 \Rightarrow \|T_k^{n-1} f - T^{n-1} f\|_K < \varepsilon$. Thus

$$\begin{aligned}\|T_k^n f - T^n f\|_C &\leq \varepsilon + \varepsilon \|T_k^{n-1} f - T^{n-1} f\| + \varepsilon \\ &\leq 2\varepsilon + 2\varepsilon \|f\| \qquad \text{for all} \quad k \geq k_1.\end{aligned}$$

■

Now we are ready to prove the continuity result; that is, $p_k \to p \Rightarrow \mu_k \to \mu$.

Proposition 3. *Suppose p, p_k are transition probabilities with associated attractive measures μ, μ_k and operators T, T_k, $k = 1, 2, \ldots$. Assume*

(i) *$p_k(x, \cdot) \to p(x, \cdot)$ for all x and $\{p_k\}$ is equicontinuous (that is, $p_k \to p$ uniformly on compact sets as discussed above);*
(ii) *$\{T^n f\}$ is equicontinuous for each $f \in C_c(X)$; and*
(iii) *for some x_0, the processes started at x_0 with transition probabilities p_k are uniformly tight; i.e., for every $\varepsilon > 0$ there exists a compact set $K \subset X$ such that $|T_k^n g(x_0)| \leq \|g\|_K + \varepsilon \|g\|$ for all $g \in C(X)$, all n, all k.*

Then $\mu_k \to \mu$.

Proof. Let $f \in C_c(X)$. Let $c = \int f \, d\mu$, $c_k = \int f \, d\mu_k$. We need to show $c_k \to c$.

Let $\varepsilon > 0$. Find K as in (iii). Since $T^n f \to c$ uniformly on compact sets by equicontinuity, we can find n_0 so that

$$\|T^{n_0} f - c\|_K < \varepsilon.$$

By the lemma, take k_0 so large that $k \geq k_0 \Rightarrow \|T_k^{n_0} f - T^{n_0} f\| < \varepsilon$, so $\|T_k^{n_0} f - c\|_K < 2\varepsilon$. Thus, for $k \geq k_0$ and $m \in \mathbf{N}$,

$$|T_k^m (T_k^{n_0} f - c)(x_0)| \leq 2\varepsilon + \varepsilon \|T_k^{n_0} f - c\|,$$

so $|T_k^{m+n_0} f(x_0) - c| \leq 2\varepsilon + 2\varepsilon \|f\|$. Letting $m \to \infty$, we get $|c_k - c| \leq 2\varepsilon + 2\varepsilon \|f\|$ for $k \geq k_0$. ■

2. Iterated Function Systems

A. The following type of random walk (i.e., discrete-time Markov process) on a metric space X is called an iterated function system (i.f.s.), generalizing considerably the definitions in [BD] and [BDEG].

Let W be the set of continuous maps from X into X. Starting at $x \in X$, choose a map $w \in W$ according to some probability distribution which is allowed to depend on x, and go to position wx. Repeat ad infinitum.

To make this description precise, assume as in Section 1 that x is separable and locally compact (this is not needed if the probability distribution of the maps is independent of position). Let W have the topology of uniform convergence on compact sets in X (i.e., the compact-open topology). W is a separable metrizable space [D, pp. 265 and 272]. The Borel probability measures $\mathscr{P}(W)$ on W have the topology of weak convergence, as discussed in Section 1.

Now suppose $\sigma: X \to \mathscr{P}(W)$ is continuous. We write $\sigma(x, \cdot)$ instead of $\sigma(x)(\cdot)$, by analogy with Section 1. Then under certain conditions, given in the next proposition,

$$p(x, B) = \int 1_B(wx)\sigma(x, dw),$$

$x \in X$, B a Borel set in X, will be a transition probability for a Markov process, the one described informally above. And $p(x, \cdot)$ will be continuous in x, so the process will be Feller, as in Section 1. By checking on indicator functions, we see that T has the form

$$Tf(x) = \int f(y)p(x, dy) = \int f(wx)\sigma(x, dw).$$

We remark that our definition of i.f.s. is so general that *every Markov Feller process on X is in fact an i.f.s.*, with $\sigma(x, \cdot)$ even supported on the constant functions for each x. To see this, define $\theta: X \to W$ by $\theta(x)(y) = x$ for all $y \in X$, $x \in X$. Now, given a transition probability p, define $\sigma(x, S) = p(x, \theta^{-1}(S))$ for S a Borel set in W. It is easy to see that this σ reproduces p when used as above, and $\sigma(x, \cdot)$ is continuous if $p(x, \cdot)$ is.

This remark has no practical application, of course; it only shows that i.f.s.'s are really just a way of viewing Markov processes, which is useful in many cases. It is the way σ is presented which determines the interest. We mention especially the simplest case when σ is independent of x and supported on finitely many maps; then

$$Tf(x) = \sum_{i=1}^{n} f(w_i x)p_i,$$

so T is an average of the composition-type operators studied in ergodic theory.

Proposition 4. *Suppose $\{\sigma(x, \cdot), x \in X\}$ is tight. Then p defined above is a transition probability for a Markov Feller process.*

In case $\sigma(x, \cdot)$ is the same for all x (so the sequence of maps chosen is an independent, identically distributed (i.i.d.) sequence), this is true when X is an arbitrary metric space and without any tightness condition on the single measure σ (although this will be automatic for any "reasonable" X, i.e., a Polish X).

Proof. Let $f \in C(X)$, let $x_n \to x_0$, and let $\varepsilon > 0$. Now

$$Tf(x_n) - Tf(x_0) = \int [f(wx_n) - f(wx_0)]\sigma(x_n, dw) + \int f(wx_0)[\sigma(x_n, dw) - \sigma(x_0, dw)].$$

Choose compact $K \subset W$ such that $\sigma(x_n, \tilde{K}) < \varepsilon$ for all n. Now K is equicontinuous [D, pp. 276, Exercise 6-1] and $\{wx_0: w \in K\}$ is compact in X, so it follows that, for sufficiently large n, $|f(wx_n) - f(wx_0)| < \varepsilon$ for all $w \in K$. Thus the first integral on the right-hand side above has magnitude $\leq \varepsilon + 2\|f\|\varepsilon$ for sufficiently large n; the second integral converges to 0 since $x_n \to x_0$ and $\sigma(x, \cdot)$ is continuous in x. Thus $Tf(x_n) \to Tf(x_0)$, so $Tf \in C(X)$.

It now follows by a standard approximation argument that, since $\int f(wx)\sigma(x, dw)$ is continuous in x for each $f \in C(X)$, $\int 1_B(wx)\sigma(x, dw)$ is Borel measurable for each Borel set B.

In the place-independent case, the result follows immediately from Lebesgue's bounded convergence theorem without any conditions on σ or X. ■

B. Next we discuss parametrization of an i.f.s., and how the continuity theorem (Proposition 3) of Section 1 translates into this language.

Let $l \in \mathbf{N}$, and let $\varphi: \mathbf{R}^l \to W$ be continuous. Suppose for each $x \in X$, $\lambda(x, \cdot) \in \mathcal{P}(\mathbf{R}^l)$ such that $x \to \lambda(x, \cdot)$ is continuous, and $\{\lambda(x, \cdot), x \in X\}$ is tight. Then defining $\sigma(x, S) = \lambda(x, \varphi^{-1}(S))$, we obviously get an i.f.s. as above. We think of $\mathbf{R}^l$ as *parameter space.* Obviously, if (ii) and (iii) of Proposition 3 are satisfied, and if $\lambda_k \to \lambda$ uniformly on compact sets, then $\sigma_k \to \sigma$ uniformly on compact sets also, so $\mu_k \to \mu$, by Proposition 3.

As an example, consider the case of $\lambda(x, \cdot)$ supported on the affine maps on $X = \mathbf{R}^n$. An affine map $wx = ax + b$ is parametrized by $l = n^2 + n$ numbers corresponding to the entries in the matrix a and translation vector b. Suppose we further restrict to the case of a fixed finite number N of affine maps. Then λ has the form

$$\lambda(x, \cdot) = \sum_{j=1}^{N} p_j(x)\delta_{t_j},$$

where $t_j \in \mathbf{R}^l$ and $\sum p_j(x) = 1$, with each p_j continuous and nonnegative. If $\lambda_k(x, \cdot) = \sum_{j=1}^{N} p_{j,k}(x)\delta_{t_{j,k}}$, then $\lambda_k \to \lambda$ uniformly on compact sets $\Leftrightarrow p_{j,k} \to p_j$ uniformly on compact sets and $t_{j,k} \to t_j$ for each j.

It follows from [BDEG] that if $\lambda_k(x, \cdot)$ are supported on maps with an *average* contractivity < 1, uniform over x and k, and if the probability of contraction is bounded away from 0 uniformly over x and k, and further that the $p_{j,k}$ have modulus of continuity satisfying Dini's condition (slightly weaker than Holder-continuity), then the other hypotheses of Proposition 3 will be satisfied.

In the next section we consider a setup with $\sigma(x, \cdot) = \sigma(\cdot)$ independent of x, supported on affine maps, not necessarily finitely many. Then $\lambda_k \to \lambda$ means just ordinary weak convergence in $\mathcal{P}(\mathbf{R}^l)$. It is clear that norm-convergence would have been the wrong thing to consider for this problem, since $\|\delta_{t_k} - \delta_t\| \not\to 0$ even though $t_k \to t$ in $\mathbf{R}^l$.

3. Homogeneous Affine Iterated Functions Systems: Solution to Inverse Problem by Fourier Transform, and Cumulants; Finite Approximation; Nature of the Invariant Measure

A. Consider an i.f.s. as in Section 2, with $X = \mathbf{R}^n$, and $\sigma(\cdot)$ *not dependent on position* and supported on the affine maps. For an affine map w, write $wx = ax + b$, with linear part a. Suppose the process has an attractive invariant measure μ. For affine maps, the Fourier transform of the invariant measure connects nicely with σ. Define

$$\hat{\mu}(t) = \int e^{it\cdot x}\mu(dx), \qquad t \in \mathbf{R}^n.$$

Because of the invariance property $\int f\,d\mu = \int Tf\,d\mu$, taking $f(x) = e^{it\cdot x}$ we get this relation:

$$\hat{\mu}(t) = \int e^{it\cdot x}\mu(dx) = \int\int e^{it\cdot(ax+b)}\sigma(dw)\mu(dx)$$
$$= \int e^{it\cdot b}\hat{\mu}(\tilde{a}t)\sigma(dw),$$

where $\tilde{a}$ is the transpose of a.

Now consider the problem of trying to recover σ from μ (or equivalently from $\hat{\mu}$). In general, a given μ can arise from more than one σ, so this is not well defined, and even if it were, the relation above appears unsolvable for σ.

However, if the linear part a is the *same* for all w in the support of σ—we call this a *homogeneous affine i.f.s.*—then we can actually solve the relation above for σ as follows (now we consider σ to be the distribution of the translations $b \in \mathbf{R}^n$, since a is constant):

$$\hat{\mu}(t) = \hat{\mu}(\tilde{a}t)\int e^{it\cdot b}\sigma(db) = \hat{\mu}(\tilde{a}t)\hat{\sigma}(t);$$

i.e.,

$$(*) \qquad \hat{\sigma}(t) = \frac{\hat{\mu}(t)}{\hat{\mu}(\tilde{a}t)}.$$

This is the key relation. In other words,

$$\sigma = \mathcal{F}^{-1}\left(\frac{\hat{\mu}(\cdot)}{\hat{\mu}(\tilde{a}\,\cdot)}\right),$$

where $\mathcal{F}^{-1}$ indicates the inverse Fourier transform. We assume above that a is a strictly contractive linear map on $\mathbf{R}^n$, or else there would not be an attractive invariant measure. We call the common a the *scale* in a homogeneous affine i.f.s. Also, we need to make some hypothesis on the drop-off rate of σ, to get tightness for the process. This is discussed later.

Proposition 5. *Let $\mu \in \mathcal{P}(\mathbf{R}^n)$. Then μ is the attractive invariant measure for a homogeneous affine i.f.s. with strictly contractive scale a if and only if $\hat{\mu}(\cdot)/\hat{\mu}(\tilde{a}\cdot)$ is positive definite, in which case the distribution σ of the translations b of the affine maps $x \mapsto ax+b$ is given by*

$$\sigma = \mathcal{F}^{-1}\left(\frac{\hat{\mu}(\cdot)}{\hat{\mu}(\tilde{a}\cdot)}\right).$$

Proof. One direction follows immediately from the key relation above, since a Fourier transform of a positive measure is positive definite.

Conversely, assume that $\hat{\mu}(\cdot)/\hat{\mu}(\tilde{a}\cdot)$ is positive definite. It is obviously continuous at 0 since $\hat{\mu}$ is and since $\hat{\mu}(0)=1$. Thus, by Bochner's theorem, it is the Fourier transform of a positive measure σ on $\mathbf{R}^n$ which must be a probability measure since $\hat{\sigma}(0)=1$. Consider the homogeneous affine i.f.s. with scale a and translation distribution σ. If the process is started at x, the position at time n is

$$Z_n^x = w_n \cdots w_1 x = a^n x + \sum_{k=1}^{n} a^{n-k} b_k,$$

where $b_1, \ldots, b_n$ are i.i.d. with distribution σ. Let ψ_n be the Fourier transform of the distribution of Z_n^x, so by the independence of the b_k's,

$$\begin{aligned}
\psi_n(t) &= e^{it\cdot a^n x} \prod_{k=1}^{n} \hat{\sigma}(a^{n-k} t) \\
&= e^{it\cdot a^n x} \prod_{k=1}^{n} \hat{\sigma}(a^{k-1} t) \\
&= e^{it\cdot a^n x} \frac{\hat{\mu}(t)}{\hat{\mu}(\tilde{a}t)} \frac{\hat{\mu}(\tilde{a}t)}{\hat{\mu}(\tilde{a}^2 t)} \cdots \frac{\hat{\mu}(\tilde{a}^{n-1} t)}{\hat{\mu}(\tilde{a}^n t)} \\
&= e^{it\cdot a^n x} \frac{\hat{\mu}(t)}{\hat{\mu}(\tilde{a}^n t)}.
\end{aligned}$$

Since $a^n \to 0$, and $\hat{\mu}$ is continuous at 0, $\psi_n(t) \to \hat{\mu}(t)$. So by the continuity theorem for Fourier transforms, Z_n^x converges in distribution to μ, so μ is the attractive invariant measure for the i.f.s. ■

Next we discuss conditions on σ implying the existence and certain drop-off rates of the invariant measure μ, and vice versa.

It is easy to see (Exercise 22.11 in [Bi], or see [BA] for more general circumstances), that

$$\int \log^+ |b| \sigma(db) < \infty$$

is sufficient for the existence of an attractive invariant measure μ; we show that if a is nonsingular, this condition is also necessary. Together with the previous proposition, this gives the somewhat surprising result that for nonsingular a, given any probability measure μ, $\hat{\mu}(\cdot)/\hat{\mu}(\tilde{a}\,\cdot)$ cannot be the Fourier transform of a positive measure σ unless σ satisfies

$$\int \log^+|b|\sigma(db) < \infty.$$

Even when a is singular, it is easy to see that if $\int \log^+|x|\mu(dx) < \infty$, then $\int \log^+|b|\sigma(db) < \infty$. The converse is false; we will give an example where $\int \log^+|b|\sigma(db) < \infty$ but $\int \log^+|x|\mu(dx) = \infty$. However, if $\int \log^2|b|\sigma(db) < \infty$, then $\int \log^+ \log^+|x|\mu(dx) < \infty$, so the drop-off rate of σ controls that of μ to some extent. For moments, the result is nicer: for $p > 0$, $\int |b|^p\sigma(db) < \infty \Leftrightarrow \int |x|^p\mu(dx) < \infty$. These things are proved using an easy modification of a result of Boas [Bo] concerning the connection between the drop-off rate of a measure and the behavior at 0 of its Fourier transform, and Minkowski's inequality.

Proposition 6. *Let σ be the distribution of translations in a homogeneous affine i.f.s. with strictly contractive scale a. Then $\int \log^+|b|\sigma(db) < \infty$ is sufficient for the existence of an attractive invariant measure. If a is* nonsingular, *this condition is also necessary.*

Proof. Sufficiency has already been discussed. To prove necessity, assume $\int \log^+|b|\sigma(db) = \infty$, and let the process start at x, so the position at time n is given by $a^n x + \sum_{k=1}^n a^{n-k} b_k$, where the $\{b_k\}$ are i.i.d. with distribution σ (see the proof of Proposition 5). This has the same distribution as $a^n x + \sum_{k=1}^n a^{k-1} b_k$. Now a series of *independent* random variables converges in distribution iff it converges almost surely, so the terms of the series must converge to 0 almost surely. But since a is nonsingular, $\|a^{-1}\| = \alpha < \infty$. We may assume without loss of generality that $\alpha > 1$ in what follows. Now

$$|a^{k-1}b_k| \geq \left(\frac{1}{\sigma}\right)^{k-1} |b_k|,$$

so

$$P(|a^{k-1}b_k| > 1) \geq P(|b_k| > \alpha^{k-1}),$$

and

$$\begin{aligned}\sum_{k=1}^{\infty} P(|b_k| \geq \alpha^{k-1}) &= \sum_{k=1}^{\infty} P(|b| \geq \alpha^{k-1}) \\ &\geq \sum_{k=1}^{\infty} P(\log^+|b| \geq (k-1)\log\alpha \text{ and } |b| \geq 1) \\ &\geq \frac{1}{\log\alpha} \int_{|b|\geq 1} \log^+|b|\sigma(db) = \infty.\end{aligned}$$

By Borel-Cantelli, we thus have

$$P(|a^{k-1}b_k| \geq 1 \text{ i.o.}) = 1.$$

This shows the series does *not* converge in distribution. ∎

Lemma 2. *Let μ be a probability measure on $\mathbf{R}^1$, and let ϕ be its Fourier transform.*

Then

(i) $$\log^+ \log^+|x|\mu(dx) < \infty \quad \textit{iff} \int_0^{1/e} \frac{|\phi(t)-1|}{t\log(1/t)}\,dt < \infty;$$

(ii) $$\int \log^2|x|\mu(dx) < \infty \quad \textit{iff} \int_0^{1/e} \frac{\log(1/t)}{t}|\phi(t)-1|\,dt < \infty.$$

Proof. These are easily proved along the same lines as Boas [Bo] proved that

$$\int \log^+|x|\mu(dx) < \infty \quad \Leftrightarrow \quad \int_o^1 \frac{|\phi(t)-1|}{t}\,dt < \infty.$$

We prove (i); (ii) is done similarly.

Assume first that

$$\int_0^{1/e} \frac{|\phi(t)-1|}{t\log(1/t)}\,dt < \infty.$$

Now

$$\frac{2-\phi(t)-\phi(-t)}{t\log(1/t)} = \frac{2}{t\log(1/t)}\int (1-\cos tx)\mu(dx),$$

so

$$\int_0^{1/e} \frac{2-\phi(t)-\phi(-t)}{t\log(1/t)} = 2\int\int_0^{1/e} \frac{1-\cos tx}{t\log(1/t)}\,dt\mu(dx)$$

$$= \int\int_0^{|x|/e} \frac{1-\cos s}{s\log(|x|/s)}\,ds\mu(dx).$$

But there exists x_0 and $C > 0$ such that, for $|x| \geq x_0$,

$$\int_0^{|x|/e} \frac{1-\cos s}{s\log(|x|/s)}\,ds \geq C\int_1^{|x|/e} \frac{1}{s\log(|x|/s)}\,ds = C\log\log|x|.$$

(This follows from monotonicity of $s\log(|x|/s)$ over the interval.) Thus

$$\int_0^{1/e} \frac{|1-\phi(t)|}{t\log(1/t)}\,dt \geq \tfrac{1}{2}\int_0^{1/e} \frac{2-\phi(t)-\phi(-t)}{t\log(1/t)}\,dt$$

$$\geq \int_{|x|\geq x_0} \log\log x_{\mu}(dx),$$

which proves one direction.

Going the other way, since

$$|\phi(t)-1| \leq \int 2|\sin \tfrac{1}{2}xt|\mu(dx),$$

we have

$$\int_0^{1/e} \frac{|\phi(t)-1|}{t\log(1/t)}\,dt \le 2\int\int_0^{1/e} \frac{|\sin\frac{1}{2}xt|}{t\log(1/t)}\,dt\mu(dx)$$
$$\le 2\int\int_0^{|x|/e} \frac{|\sin\frac{1}{2}s|}{s\log|x|/s}\,ds\mu(dx).$$

But, for $|x| \ge e$,

$$2\int_0^1 \frac{|\sin\frac{1}{2}s|}{s\log|x|/s}\,ds \le \int_0^1 \frac{1}{\log|x|/s}\,ds = |x|\int_{\log|x|}^{\infty} \frac{e^{-y}}{y}\,dy$$
$$\le |x|\int_{\log|x|}^{\infty} e^{-y}\,dy = 1,$$

and

$$2\int_1^{|x|/e} \frac{1}{s\log|x|/s}\,dx = 2\log\log|x|.$$

For $|x| < e$,

$$2\int_0^{1/e} \frac{|\sin\frac{1}{2}xt|}{t\log(1/t)}\,dt \le e\int_0^{1/e} \frac{1}{\log(1/t)}\,dt \le e.$$

Thus

$$\int_0^{1/e} \frac{|\phi(t)-1|}{t\log(1/t)}\,dt \le 1+e+\int \log^+\log^+|x|\mu(dx). \qquad \blacksquare$$

Proposition 7. *Let σ be the distribution of translations for a homogeneous affine i.f.s. with a strictly contractive scale a, and assume that an attractive invariant measure μ exists.*

(i) *If $\int \log^2|b|\sigma(db) < \infty$, then $\int \log^+\log^+|x|\mu(dx) < \infty$. If $\int \log^+|x|\mu(dx) < \infty$, then $\int \log^+|b|\sigma(db) < \infty$ (recall this holds automatically if a is nonsingular).*
(ii) *For $p > 0$, $\int |b|^p\sigma(db) < \infty \Leftrightarrow \int |x|^p\mu(dx) < \infty$.*

Proof. As in the proofs of Propositions 5 and 6, letting $S_n = \sum_{k=1}^n a^{k-1}b_k$ where the $\{b_k\}$ are i.i.d. with distribution σ, S_n converges in distribution to μ. Because of convergence, choose λ so that

$$P\left(\left|\sum_{k=2}^n a^{k-1}b_k\right| > \lambda\right) < \tfrac{1}{2}$$

for all n. Because of independence, for all n,

$$P(|S_n| \ge c) \ge P\left(|b_1| \ge c+\lambda \text{ and } \left|\sum_{k=2}^n a^{k-1}b_k\right| \le \lambda\right)$$
$$\ge \tfrac{1}{2}P(|b_1| \ge c+\lambda).$$

Thus, for all c,

$$\mu\{x: |x| \geq c\} \geq \tfrac{1}{2}\sigma\{b: |b| \geq c+\lambda\}.$$

From this it is apparent that

$$\int \log^+|x|\mu(dx) < \infty \left(\text{resp.} \int |x|^p \mu(dx) < \infty\right) \Rightarrow$$

$$\int \log^+|b|\sigma(db) < \infty \left(\text{resp.} \int |b|^p \sigma(db) < \infty\right).$$

Conversely, assume $\int \log^2|b|\sigma(db) < \infty$. It is easy to see that it is enough to consider the one-dimensional case only, so assume $X = \mathbf{R}^1$. By Lemma 2, we have

$$\int_0^{1/e} \frac{\log 1/t}{t} |\hat{\sigma}(t) - 1| \, dt < \infty.$$

Since $\hat{\sigma}(t) = \hat{\mu}(t)/\hat{\mu}(at)$ and $|\hat{\mu}(at)| \leq 1$,

$$\int_0^{1/e} \frac{\log 1/t}{t} |\hat{\mu}(t) - \hat{\mu}(at)| \, dt < \infty.$$

Now

$$\hat{\mu}(t) - 1 = \sum_{k=0}^{\infty} \hat{\mu}(a^k t) - \hat{\mu}(a^{k+1} t)$$

since $\hat{\mu}(a^n t) \to 1$, so

$$\begin{aligned}
\int_0^{1/e} \frac{|\hat{\mu}(t) - 1|}{t \log(1/t)} dt &\leq \sum_{k=0}^{\infty} \int_0^{1/e} \frac{|\hat{\mu}(a^k t) - \hat{\mu}(a^{k+1} t)|}{t \log(1/t)} dt \\
&= \sum_{k=0}^{\infty} \int_0^{a^k/e} \frac{|\hat{\mu}(s) - \hat{\mu}(as)|}{s \log(a^k/s)} ds \\
&\leq \sum_{k=0}^{\infty} \int_0^{a^k/e} \frac{\log(1/s)}{s} |\hat{\mu}(s) - \hat{\mu}(as)| \frac{1}{\log(1/s)} ds
\end{aligned}$$

since $\log(a^k/s) \geq 1$.

Sublemma. *If g is a nonnegative, Lebesgue-integrable function on $[0, 1/e]$ and $0 < a < 1$, then*

$$\sum_{k=0}^{\infty} \int_0^{a^k/e} \frac{g(s)}{\log 1/s} ds < \infty.$$

Proof. By the integral test, we only need observe that

$$\begin{aligned}
\int_0^{\infty} \int_0^{a^k/e} \frac{g(s)}{\log(1/s)} ds \, dk &= \int_0^{1/e} \frac{g(s)}{\log(1/s)} \int_0^{(1+\log s)/\log a} dk \, ds \\
&\leq \int_0^{1/e} g(s) \left(\frac{1}{\log a}\right) ds < \infty.
\end{aligned}$$

■

Returning to the proof of the proposition, apply the sublemma with

$$g(s)=\frac{\log(1/s)}{s}|\hat{\mu}(s)-\hat{\mu}(as)|$$

to the above to yield

$$\int_0^{1/e}\frac{|\hat{\mu}(t)-1|}{t\log(1/t)}\,dt<\infty,$$

so by Lemma 2

$$\int \log^+\log^+|x|\mu(dx)<\infty.$$

Finally, assume $\int|b|^p\sigma(db)<\infty$. Since $S_n=\sum_{k=1}^n a^{k-1}b_k$, we have, for $p\geq 1$, by Minkowski's inequality,

$$\begin{aligned}(\mathbf{E}|S_n|^p)^{1/p}&\leq\sum_{k=1}^{\infty} r^{k-1}\left(\int|b|^p\sigma(db)\right)^{1/p}\\&=\frac{1}{1-r}\left(\int|b|^p\sigma(db)\right)^{1/p},\end{aligned}$$

where $r=\|a\|<1$. If, however, $p<1$, we have directly $|S_n|^p\leq\sum_{k=1}^n r^{p(k-1)}|b_k|^p$, so $\mathbf{E}(|S_n|^p)\leq(1/(1-r^p))\int|b|^p\sigma(db)$. In either case, since S_n converges in distribution to μ,

$$\int|x|^p\mu(dx)<\infty. \qquad ■$$

As an example where $\int\log^+|b|\sigma(db)<\infty$ but $\int\log^+|x|\mu(dx)=\infty$, let σ (on $\mathbf{R}^1$) have density $g(b)=c/(b\log^3 b)$ for $b\geq e$. Let $0<a<1$. We show that for sufficiently large n,

$$E\log(S_{n+1})\geq E\log(S_n)+\frac{K}{n},$$

where $K>0$, so $E(\log S_n)\to\infty$.

Now $S_{n+1}=S_n+a^n b_{n+1}$, so

$$\log(S_{n+1})=\log S_n+\log\left(1+\frac{a^n b_{n+1}}{S_n}\right).$$

Since b_{n+1} is independent of s_n,

$$\begin{aligned}\mathbf{E}(\log S_{n+1}|S_n=y)&=y+\int_e^{\infty}\log\left(1+\frac{a^n b}{y}\right)g(b)\,db\\&\geq\log y+\int_{a^{3n}b>y}\log(a^n b/y)g(b)\,db\\&=\log y+\int_{a^{3n}b>y}\log(a^n b)\frac{c}{b\log^3 b}\,db\\&\quad-\log y\sigma\{b\colon a^{3n}b>y\}.\end{aligned}$$

But $\log a^n b \geq \frac{2}{3}\log b$ if $a^{3n}b > 1$, so, since $y \geq e$,

$$\mathbf{E}(\log S_{n+1} | S_n = y) \geq \log y + \tfrac{2}{3}c \int_{a^{3n}b>y} \frac{1}{b\log^2 b}\,db$$

$$-\log y \int_{a^{3n}b>y} \frac{c}{b\log^3 b}\,db$$

$$= \log y + \tfrac{2}{3}c\frac{1}{\log(y/a^{3n})} - \log y \frac{c}{2\log^2(y/a^{3n})}$$

$$\geq \log y + \frac{c}{b}\frac{1}{\log(y/a^{3n})}.$$

Now $\log y \leq 3n\log(1/a)$ for $y \leq (1/a)^{3n}$, so

$$\mathbf{E}(\log S_{n+1}) \geq \mathbf{E}(\log S_n) + \frac{c}{6}\frac{1}{2(3n\log(1/a))}P\left(S_n \leq \left(\frac{1}{a}\right)^{3n}\right).$$

But since S_n converges in distribution, $P(S_n \leq (1/a)^{3n}) \geq \frac{1}{2}$ for large n, so $\mathbf{E}(\log S_{n+1}) \geq \mathbf{E}(\log S_n) + K/n$, which completes the demonstration.

The example could be modified to show that given any unbounded positive increasing function θ, there is a σ with $\int \log^+|b|\sigma(db) < \infty$ but $\int \theta(|x|)\mu(dx) = \infty$, so there is no hope for a better result.

Next, we make some simple observations about the compactly supported case. Obviously, if σ is supported on $\{|b| \leq M\}$ (that is, has support contained in this set), then μ is supported on $\{|x| \leq (1/(1-r))M\}$, where $r = \|a\| < 1$ (recall $\sum_{k=1}^{n} a^{k-1}b_k$ converges in distribution to μ). Conversely,

Proposition 8. *Let σ be the translation distribution for a homogeneous affine i.f.s. of scale a, where $\|a\| = r < 1$, with attractive invariant measure μ.*

(i) *If μ is supported on $\{|x| \leq K\}$, then σ is supported on $\{|b| \leq (1+r)K\}$.*
(ii) *If $X = \mathbf{R}^1$, $a > 0$, and μ is supported on $[0, K]$, then σ is supported on $[0, K(1-r)]$.*

Proof. (i) Let $\varepsilon > 0$, and choose $0 \leq f \leq 1$ continuous such that $f \equiv 0$ on $\{|x| \leq K\}$, $f \equiv 1$ on $\{|x| \geq K + \varepsilon\}$. Then

$$0 = \int f\,d\mu = \int Tf\,d\mu = \int\int f(ax+b)\sigma(db)\mu(dx)$$

$$\geq \int_{|b| \geq K(1+r)+\varepsilon} \int f(ax+b)\mu(dx)\sigma(db)$$

$$\geq \int_{|b| \geq K(1+r)+\varepsilon} \int \mu\{x: |ax+b| \geq K+\varepsilon\}\sigma(db).$$

But $|x| \le K$ and $|b| \ge K(1+r)+\varepsilon \Rightarrow |ax+b| \ge K+\varepsilon$, so

$$0 \ge \int_{|b| \ge K(1+r)+\varepsilon} \mu\{x: |x| \le K\}\sigma(db) = \sigma\{b: |b| \ge K(1+r)+\varepsilon\}.$$

This proves (i).

For (ii), assume without loss of generality that K is a point of support for μ. Let $\varepsilon > 0$, and let $0 \le f \le 1$ be continuous such that $f \equiv 0$ on $[0, K]$ and $f \equiv 1$ on $[K+\varepsilon, \infty)$. Then

$$0 = \int f\,d\mu = \int Tf\,d\mu = \int\int f(ax+b)\sigma(db)\mu(dx)$$

$$\ge \int_{b \ge K(1-a)+2\varepsilon} \mu\{x: ax+b \ge K+\varepsilon\}\sigma(db).$$

But $x \ge K - \varepsilon/a$ and $b \ge K(1-a)+2\varepsilon \Rightarrow ax+b \ge K+\varepsilon$, so

$$0 \ge \int_{b \ge K(1-a)+2\varepsilon} \mu\left\{x: x \ge K - \frac{\varepsilon}{a}\right\}\sigma(db) \Rightarrow \sigma\{b: b \ge K(1-a)+2\varepsilon\} = 0$$

since K is a point of support of μ. ■

Next, we consider the problem of approximating μ, when $\hat{\mu}(\cdot)/\hat{\mu}(\tilde{a}\,\cdot)$ is positive definite, by invariant measures for homogeneous affine i.f.s.'s with linear part a and σ supported on only *finitely* many points (that is, an i.f.s. with only finitely many maps). Proposition 3 will make this simple.

Proposition 9. *A measure $\mu \in \mathcal{P}(\mathbf{R}^n)$ is the limit of invariant measures for finite homogeneous affine i.f.s.'s with linear part a if and only if $\hat{\mu}(\cdot)/\hat{\mu}(\tilde{a}\,\cdot)$ is positive definite.*

Proof. Assume first that $\mu_k \to \mu$, where μ_k is the invariant measure for a homogeneous affine i.f.s. with linear part a and finitely supported translation distributions σ_k. Then

$$\hat{\sigma}_k(t) = \frac{\hat{\mu}_k(t)}{\hat{\mu}_k(\tilde{a}t)},$$

so by the continuity theorem for Fourier transforms

$$\hat{\sigma}_k(t) \to \frac{\hat{\mu}(t)}{\hat{\mu}(\tilde{a}t)}$$

which is continuous at 0, so $\hat{\mu}(t)/\hat{\mu}(\tilde{a}t)$ is the Fourier transform of a probability measure, hence positive definite.

Conversely, assume $\hat{\mu}(\cdot)/\hat{\mu}(\tilde{a}\,\cdot)$ is positive definite. By Proposition 5, μ is the invariant measure for a homogeneous affine i.f.s. with linear part a and distribution of translations σ. By Proposition 3, we need only find $\sigma_k \in \mathcal{P}(\mathbf{R}^n)$, $k = 1, 2, \ldots,$ each supported on finitely many points, such that $\sigma_k \to \sigma$ (see the discussion in Section 2B). But this can obviously be done by taking succesively finer finite partitions $\{B_{k,i},\ i=1,\ldots,n_k\}$ of $\mathbf{R}^n$ and setting

$$\sigma_k = \sum_{i=1}^{n_k} \sigma(B_{k,i})\delta_{x_{k,i}},$$

where $x_{k,i} \in B_{k,i}$. ■

Remark. This is not necessarily a *good* way to approximate, of course.

B. Now we discuss the cumulants/moments method for solving the inverse problem for a homogeneous affine i.f.s.

The key equation connecting σ with the invariant measure μ was given in Subsection A in terms of the Fourier transform as

$$\hat{\sigma}(t) = \frac{\hat{\mu}(t)}{\hat{\mu}(\tilde{a}t)}, \qquad t \in \mathbf{R}^n.$$

Obviously this same equation holds for the Laplace transform as well, when it exists: just replace it by t. *We assume from now on that μ has a drop-off rate sufficient to guarantee existence of a Laplace transform* in some neighborhood of 0.

Let us denote the Laplace transform of μ by

$$\overset{>}{\mu}(t) = \int e^{tx}\mu(dx), \qquad t \in \mathbf{R}^n.$$

(We rotate $\hat{}$ by $\pi/2$ to $^{>}$ since ti rotates clockwise by $\pi/2$ to become t!.) Thus

$$\overset{>}{\sigma}(t) = \frac{\overset{>}{\mu}(t)}{\overset{>}{\mu}(\tilde{a}t)},$$

so σ could be recovered as the inverse Laplace transform of $\overset{>}{\mu}(\cdot)/\overset{>}{\mu}(\tilde{a}\,\cdot)$.

Instead, we want to give the very simple relation that exists between the cumulants (see below for definition) of σ and μ, which could be exploited to solve for σ in terms of μ. The moments are easily obtained from the cumulants and vice versa, as discussed below. In order not to obscure the idea with horrendous notation, *we assume for the remainder of Subsection B that $X = \mathbf{R}^1$.* Thus a becomes simply a number, with $0<|a|<1$. This method could also be used in higher dimensions, but we do not discuss it.

Let $\psi(t) = \log \overset{>}{\sigma}(t)$ and $\phi(t) = \log \overset{>}{\mu}(t)$. These are sometimes called the *cumulant-generating functions*, in analogy with the Laplace transform being called the moment-generating function. The *cumulants* of μ are the numbers $\phi^{(k)}(0)$, $k = 1, 2, \ldots$.

The key relation, in terms of ψ and ϕ, is

$$\psi(t) = \phi(t) - \phi(at).$$

Differentiating and setting t to 0, we get the following easily computed relationship between cumulants:

Proposition 10. *For a homogeneous affine i.f.s. on* $\mathbf{R}^1$ *the cumulants* $\psi^{(k)}(0)$ *of* σ *are related to the cumulants* $\phi^{(k)}(0)$ *of* μ *by*

$$\psi^{(k)}(0) = (1 - a^k)\phi^{(k)}(0).$$

To make use of this, we need to be able to compute the cumulants from the moments and vice versa efficiently, since the moments can be calculated numerically from the measure, and, with some difficulty, vice versa.

We now give a simple recurrence scheme for calculating cumulants efficiently from moments, and vice versa. Surely this is known to others, but we could find no mention of it in any probability/statistics references, where cumulants are discussed. Other less-efficiently computable relations are usually given.

Since the moments/cumulants are simply power series coefficients of the moment-generating function/cumulant-generating function, multiplied by appropriate factorials, it is enough to discuss the relation between the power series coefficients.

Lemma 3. *Let* $f(t) = \sum_{j=0}^{\infty} b_j t^j$, $b_0 = 1$, *converge in some neighborhood of* 0, *and assume* $f(t) > 0$. *Let*

$$g(t) = \log f(t) = \sum_{j=0}^{\infty} a_j t^j,$$

so $a_0 = 0$ *and* $a_1 = b_1$. *Then the following recurrence formula holds*:

$$\sum_{j=1}^{k-1} j a_j b_{k-j} + k a_k = k b_k, \qquad k \geq 2.$$

From this, $a_1, \ldots, a_n$ *may be computed from* $b_1, \ldots, b_n$ *in about* $n^2/2$ *operations, and the same for computing the* b_j*'s from the* a_j*'s.*

Proof. $g'(t) = f'(t)/f(t)$; that is,

$$(\sum b_j t^j)(\sum j a_j t^{j-1}) = \sum j b_j t^{j-1}.$$

By the usual method for multiplying power series, we get, by equating coefficients,

$$\sum_{j=1}^{k} j a_j b_{k-j} = k b_k.$$

Since $b_0 = 1$, we get the desired relation.

Note that if the b_j are known and $j a_j$ has been computed for $j < k$, we can compute $k a_k$ in about k more operations, hence the $n^2/2$ formula. Similarly for the opposite direction. ■

Acknowledgments. John H. Elton was supported in part by DARPA Applied Mathematical Sciences Program. The authors would like to thank Stephen Demko for many helpful discussions and suggestions, and in particular for suggesting the question answered in Proposition 9.

References

[BDEG] M. F. Barnsley, S. Demko, J. Elton, J. Geronimo (1988): *Invariant measures for Markov processes arising from function iteration with place-dependent probabilities.* Ann. Inst. H. Poincaré, **24**.

[BD] M. F. Barnsley, S. Demko (1985): *Iterated function systems and the global construction of fractals.* Proc. Roy. Soc. London Ser. A, **399**:243–275.

[BE] M. F. Barnsley, J. Elton (1988): *A new class of Markov processes for image encoding.* Adv. in Appl. Probab., **20**:14–32.

[BS] M. F. Barnsley, A. Sloan (preprint): *Image compression.*

[BA] M. Berger Y. Amit (preprint): *Products of random affine maps.*

[Bi] P. Billingsley (1986): Probability and Measure, 2nd ed. New York: Wiley.

[Bo] R. P. Boas (1967): *Lipschitz behavior and integrability of characteristic functions.* Ann. of Math. Statist., **38**:32–36.

[Dg] J. Dugundji (1966): Topology. Boston: Allyn and Bacon.

J. H. Elton
School of Mathematics
Georgia Institute of Technology
Atlanta
Georgia 30332
U.S.A.

Zheng Yan
School of Mathematics
Georgia Institute of Technology
Atlanta
Georgia 30332
U.S.A.

Constr. Approx. (1989) 5: 89–98

CONSTRUCTIVE APPROXIMATION

An Exact Formula for the Measure Dimensions Associated with a Class of Piecewise Linear Maps

J. S. Geronimo and D. P. Hardin

Abstract. An exact formula for the various measure dimensions of attractors associated with contracting similitudes is given. An example is constructed showing that for more general affine maps the various measure dimensions are not always equal.

1. Introduction

Recently there has been much interest in sets having fractional dimension. Mandelbrot [M] pioneered their use to model a wide variety of scientific phenomena and gave the name fractal to those sets whose Hausdorff–Besicovitch dimension differs from their topological dimension. These types of sets can exhibit a high degree of complexity which is maintained upon magnification and which appears to be a characteristic of many sets associated with natural phenomena, i.e., Brownian motion, turbulent flow, leaves. A general theoretical framework has been established [H], [BD], [BDEG], [B], [DS] on the construction of certain classes of fractals using iteration theory. In this case (see below) using N maps, say, w_i, $i=1,\ldots,N$, we construct the compact fractal set S satisfying $S=\bigcup_i w_i(S)$. Associated with maps w_i are probabilities $p_i>0$, $i=1,\ldots,N$, $\sum p_i=1$ determining the frequency with which the maps w_i are to be chosen. These probabilities give rise to an invariant measure on S.

Recent work [DS], [DHN] in the areas of computer graphics and data compression have exploited the fact that fractal sets, generated by the methods mentioned above, can be used to approximate natural objects of interest [H], [BEHL]. In particular, fractals and their measures that are associated with affine maps and their probabilities have been shown to be useful in approximating complicated natural sets and measures. Another area of approximation theory where affine maps have proved to be useful is in approximating continuous but nowhere differentiable functions [B].

Date received: February 2, 1987. Date revised: February 11, 1988. Communicated by Michael F. Barnsley.
AMS classification: 26A, 28A, 41A, 44A.
Key words and phrases: Hausdorff dimension, Lyapunov dimension, Similitudes.

A question that has been raised by the above studies is how do we determine which affine maps and probabilities are we to use to approximate the set of interest? In other words, how are we to parametrize fractal sets. One way is to use the various dimensions associated with these sets, i.e., the Hausdorff-Besicovitch dimension [M] $H(S)$, the box dimension $C(S)$, the Renyi information dimension $R(S)$, and the analogous measure dimensions $H(\mu)$, $C(\mu)$, and $R(\mu)$ [Y], [FOY], [L]. The object of this paper is to give an exact expression for the above measure dimensions in the case when the affine maps are contracting similitudes with nonzero constant probabilities associated with each map.

This problem was posed by Farmer, Ott, and Yorke [FOY] who conjectured that typically the measure dimensions are the same. Using combinatorial arguments they showed that the conjecture is true in the case of dynamical systems constructed from two one-dimensional similitudes with equal contraction factors. Later Young [Y] showed that the conjecture is true in the case of $C^{1+\alpha}$ diffeomorphisms of compact 2-manifolds. Here we consider the case of any finite number of finite-dimensional similitudes and we show that the conjecture of [FOY] is true.

We proceed as follows: in Section 2 we define all the necessary quantities and set up the machinery necessary to prove the main theorem. In Section 3 we prove the main theorem. Then, in Section 4, we present an example showing that the measure dimensions are not always the same.

2. Preliminaries

Let $K\subseteq\mathbf{R}^n$ be compact and let $w_i: K \circlearrowleft$, $i=1,2,\ldots,N$, be continuous and contractive. Then $\{K, w_i\colon i=1,\ldots,N\}$ is called a hyperbolic iterated function system (h.i.f.s.) on K. The attractor $A\subset K$ associated with the above h.i.f.s. is the unique compact set satisfying $A=\bigcup_i w_i(A)$ [BD], [H]. Given probabilities p_i, $i=1,\ldots,N$, $p_i>0$, and $\sum_{i=1}^N p_i=1$ there exists a unique invariant measure μ supported on A having the property that $\forall f\in C(A)$

$$\int_A f\,d\mu=\sum_{i=1}^N p_i\int_A f\circ w_i\,d\mu. \tag{2.1}$$

This measure μ is the so-called p-balanced measure [BD]. Let $\Omega=\{\mathbf{I}=(i_1,i_2,\ldots)\colon$ i_j an integer, $1\le i_j\le N$ for each j. Let P be the product probability measure on Ω with $\{p_{i_j}, i_j=1,2,\ldots,N\}\}$ being the probability distributed on each factor. In this case it is well known [BD, Theorems 3 and 4], [H] that

$$\lim_{n\to\infty} w_{i_1}\circ w_{i_2}\circ\cdots\circ w_{i_n}(x)=\varphi(\mathbf{i})$$

exists and is independent of x and μ has the representation $d\mu=dP\circ\varphi^{-1}$.

The Hausdorff-Besicovitch dimension $H(K)$ of K is defined to be

$$H(K)=\inf\left\{\alpha\colon \lim_{\varepsilon\to 0}\inf_{\rho}\sum_i(\operatorname{diam} S_i)^\alpha\right\},$$

where ρ is a countable covering of K by closed spheres of diameter less than or equal to ε. The Hausdorff–Besicovitch dimension of a Borel probability measure on K is given by [Y]

$$H(\mu) = \inf_{\substack{Y \subset K \\ \mu(Y)=1}} H(Y).$$

If K is the support of μ we see that $H(\mu) \leq H(K)$.

The box dimension (capacity) of a bounded set B is given by

$$C(B) = \limsup_{\varepsilon \to 0} \frac{\log N(\varepsilon)}{\log(1/\varepsilon)},$$

where $N(\varepsilon)$ is the least number of ε-balls needed to cover B.

The box dimension of a Borel probability measure μ, $C(\mu)$ is defined in the following way. For $\varepsilon, \delta > 0$, let $N(\varepsilon, \delta)$ be the minimum number of ε-balls needed to cover a set of μ-measure $\geq 1 - \delta$. Then

$$C(\mu) = \sup_{\delta \to 0} \limsup_{\varepsilon \to 0} \frac{\log N(\varepsilon, \delta)}{\log(1/\varepsilon)}.$$

Let M be a Borel subset of $\mathbf{R}^m$ and let ν be a Borel probability measure on M. Denote by B_ν the σ-algebra of all ν-measurable subsets of M. Let $W: M \circlearrowleft$ be a ν-measurable map. If W is ν-invariant, i.e., $\nu(W^{-1}(E)) = \nu(E)$ for all $E \subset B_\nu$, then the quadruple (M, B_ν, W, ν) is called a dynamical system. The above h.i.f.s. can be made into a dynamical system using the following construction [FOY], [Mas], [P]. Let $M = [0, 1] \times K$, $\nu = \mu \cdot m$, where m is uniform Lebesgue measure, $B_\nu = B_\mu \times B_m$, and

$$W(y, x) = \left(\frac{y - r_i}{p_i}, w_i(x) \right) \qquad \text{if} \quad (y, x) \in [r_{i-1}, r_i] \times K.$$

Here $r_i = \sum_{j=1}^{i} p_j$ and $r_0 = 0$.

Associated with a dynamical system is the Lyapunov dimension. Although this dimension can be defined under more general conditions, we will consider the case when the maps w_i, $i = 1, \ldots, N$, mentioned above are affine maps with $0 < s_i < 1$, $i = 1, \ldots, N$, their respective contraction factors. In this case the derivative TW of W exists for almost every $(y, x) \in [0, 1] \times K$. Let $T^{\circ m}(W)$, $m = 1, 2, \ldots$, be the tangent map induced by the mth iterate of W. Since W is linear, TW is a constant and $T^{\circ m}(W) = (TW)^m$. Let $l_1(m) \geq l_2(m) \geq \cdots \geq l_{N+1}(m)$ be the eigenvalues of $(T^{\circ m}(W))^* T^{\circ m}(W)$. Then the Lyapunov exponents of $W, \lambda_1, \lambda_2, \ldots, \lambda_{N+1}$, are defined to be

$$\lambda_i = \lim \frac{1}{2m} \ln l_i(m)$$

which exists for ν almost all $(y, x) \in [0, 1] \times K$. Let

$$k = \max\{i \in \{1, \ldots, N\}: \lambda_1 + \lambda_2 + \cdots + \lambda_i > 0\}.$$

If no such k exists then the Lyapunov dimension $\Lambda(\nu)$ of ν is defined to be zero. If $k = N$, then $\Lambda(\nu)$ is defined to be N. Finally, if $1 \le k < N$, then

$$\Lambda(\nu) = k + \frac{\lambda_1 + \cdots + \lambda_k}{|\lambda_{k+1}|}.$$

3. The Main Result

A transformation $T: R^n \to R^n$ is called a similitude if $d(T(x), T(y)) = sd(x, y)$ for all $x, y \in \mathbf{R}^n$. Here we show, using some ideas of [H], that in the case of contracting similitudes all the measure dimensions are the same. We will need the following.

Theorem 1 [Y]. *Let μ be a Borel probability measure on a compact Riemannian manifold and suppose, for μ-a.e. x,*

$$\lim_{\rho \to 0} \frac{\log(\mu(B_\rho(x)))}{\log \rho} = \alpha_0,$$

where $B_\rho(x)$ is a ball of diameter ρ centered at x. Then $H(\mu) = C(\mu) = \alpha_0$.

Theorem 2. *Let $(K, w_i: i = 1, 2, \ldots, N)$ be an h.i.f.s. with attractor A such that:*

(i) *$w_i(A) \cap w_j(A) = \emptyset$ for $i \ne j$,*
(ii) *w_i is a similitude for $i = 1, 2, \ldots, N$,*
(iii) *$0 < s_1 \le s_2 \le \cdots \le s_N < 1$.*

For a fixed probability vector $\bar{p} = (p_1, p_2, \ldots, p_N)$ let μ be the p-balanced measure. Then

$$\lim_{\rho \to 0} \frac{\log(\mu(B_\rho(x)))}{\log \rho} = D = \frac{\sum_{i=1}^{N} p_i \log p_i}{\sum_{i=1}^{N} p_i \log s_i}$$

for μ-a.e. x. Here $B_\rho(x)$ is a ball of diameter ρ centered at x.

Theorems 1 and 2 imply

Corollary 1. *With the same hypotheses as in Theorem 2,*

$$H(\mu) = C(\mu) = D.$$

We first prove

Lemma 1. *Fix $x \in A$ and let $(i_1, i_2, \ldots)$ be the code for x, i.e., $x = \lim_{n \to 0} w_{i_1} \circ w_{i_2} \circ \cdots \circ w_{i_n}(y)$. Suppose the hypotheses of Theorem 2 hold. Let $\rho > 0$ and let l be the smallest integer such that $w_{i_1} \circ w_{i_2} \circ \cdots \circ w_{i_l}(A) \subset B_\rho(x)$. Then*

$$\mu(B_\rho(x)) \ge C_1 \prod_{m=1}^{l} \left(\frac{p_{i_m}}{s_{i_m}^D}\right) \rho^D$$

for some $C_1 > 0$.

Proof. From the definition of l and the fact that $s_1 \leq s_j$ for $j = 1, 2, \ldots, N$ we find

$$\left(\prod_{m=1}^{l} s_{i_m}\right) \operatorname{diam}(A) \geq \rho s_1.$$

Consequently, from (2.1) we obtain

$$\begin{aligned} \mu(B_\rho(x)) &\geq \mu(w_{i_1} \circ w_{i_2} \circ \cdots \circ w_{i_l}(A)) = \prod_{m=1}^{l} p_{i_m} \\ &\geq \prod_{i=1}^{l} \frac{p_{i_m}}{s_{i_m}^D} \rho^D \frac{s_1^D}{(\operatorname{diam}(A))^D}, \end{aligned}$$

from which the result follows. ■

Since A is compact it follows from (i) that there exists an $\varepsilon > 0$ such that

(3.1) $$d(w_i(A), w_j(A)) \geq \varepsilon > 0, \qquad i \neq j.$$

Lemma 2. *Let $x, y \in A$ with codes $(i_1, i_2, \ldots)$ and $(j_1, j_2, \ldots)$, respectively. If $d(x, y) < \varepsilon \prod_{m=1}^{l} s_{i_m}$, then $j_m = i_m$ for $m = 1, 2, \ldots, l+1$.*

Proof. Suppose $j_m = i_m$ for $m = 1, 2, \ldots, k$ and $j_{k+1} \neq i_{k+1}$, then

$$\begin{aligned} d(x, y) &= \prod_{m=1}^{k} s_{i_m} d(w_{i_{k+1}}(x'), w_{j_{k+1}}(y')) \\ &\geq \prod_{m=1}^{k} s_{i_m} \varepsilon, \end{aligned}$$

since $x', y' \in A$ which yields a contradiction. ■

Let $\rho > 0$ and for each code $(j_1, j_2, \ldots)$ select the least q such that $\prod_{m=1}^{q} s_{j_m} \leq \rho$. Let $S(\rho)$ be the set of finite codes $(j_1, \ldots, j_q)$. Then

$$A = \bigcup_{(j_1, \ldots, j_q) \in S(\rho)} w_{j_1} \circ w_{j_2} \circ \cdots \circ w_{j_q}(A).$$

Lemma 3. *There exists $C_2 > 0$ such that*

$$\mu(B_\rho(x)) \leq C_2 \left(\prod_{m=1}^{q^*} \frac{p_{i_m}}{s_{i_m}^D}\right) \rho^D,$$

where q^* is the least q such that $\prod_{m=1}^{q} s_{i_m} \leq \rho$.

Proof. By the definition of q^* we have $s_1 \rho \leq \prod_{m=1}^{q^*} s_{i_m} \leq \rho$. Let t be the smallest integer such that $s_N^t \leq s_1 \varepsilon$ where ε is given by (3.1) (recall $s_1 \leq s_2 \leq \cdots \leq s_N$). Taking ρ small enough to ensure $q^* > t$, we find

$$\rho \leq \frac{1}{s_1} \prod_{m=1}^{q^*} s_{i_m} \leq \prod_{m=1}^{q^*-t} s_{i_m} \frac{s_N^t}{s_1} \leq \prod_{m=1}^{q^*-t} s_{i_m} \varepsilon.$$

Thus if $y \in B_\rho(x)$ with code $(j_1, j_2, \ldots)$, by Lemma 2 we must have $j_m = i_m$, $m = 1, 2, \ldots, q^* - t + 1$. Let r be the smallest integer such that $s_N^r \leq s_1^t$, then there are at most N^r codes in $S(\rho)$ such that $w_{j_1} \circ w_{j_2} \circ \cdots \circ w_{j_q}(A) \cap B_\rho(x) \neq \varnothing$. Furthermore, since $j_m = i_m$, $m = 1, \ldots, q^* - t + 1$, we have

$$\mu(w_{j_i} \circ \cdots \circ w_{j_q}(A)) \leq \prod_{m=1}^{q^*-t+1} p_{i_m}.$$

Consequently,

$$\mu(B_\rho(x)) \leq N^r \prod_{m=1}^{q^*-t+1} p_{i_m}$$

$$\leq \frac{N^r}{s_1^{D(t-1)}} \left(\prod_{i=1}^{q^*-t+1} \frac{p_{i_m}}{s_{i_m}^D} \right) \rho^D.$$

Proof of Theorem 2. Note that the l in Lemma 1 and the q^* in Lemma 3 tend to infinity as ρ decreases to zero. By the law of large numbers for μ-a.e. $x \in A$ we have

$$\lim_{n\to\infty} \log\left(\prod_{m=1}^{n} \frac{p_{i_m}}{s_{i_m}^D} \right) = \lim_{n\to 0} \log\left(\prod_{i=1}^{N} \frac{p_i^{p_i n}}{s_i^{p_i D n}} \right) = 0.$$

This together with Lemmas 1 and 3 imply

$$\lim_{\rho\to 0} \frac{\log(\mu(B_\rho(x)))}{\log \rho} = D. \qquad \blacksquare$$

Remarks. (1) Consider the dynamical system constructed as in the previous section using the maps w_i: $i = 1, \ldots, N$ with invariant measure $\nu = \mu \times m$. Under these circumstances we have $H(\nu) = H(\mu) + 1 = C(\mu) + 1 = C(\nu)$. Furthermore, note that in this case there are only two distinct Lyapunov exponents, $\lambda_1 = -\sum p_i \log(p_i) > 0$, and

$$\lambda_i = \sum_{j=1}^{N} p_j \log s_j = \lambda_2, \qquad i = 3, 4, \ldots, n+1.$$

We note that the condition $w_i(A) \cap w_j(A) = \varnothing$ implies that $k = \max\{j: \lambda_1 + \lambda_2 + \cdots + \lambda_k > 0\} < n$. Thus the Lyapunov dimension

$$\Lambda(\nu) = k + \frac{(k-1)\lambda_2 + \lambda_1}{-\lambda_2} = 1 - \frac{\lambda_1}{\lambda_2}$$

$$= 1 + \frac{\sum_{i=1}^{N} p_i \log p_i}{\sum_{i=1}^{N} p_i \log s_i} = H(\nu) = C(\nu).$$

For this class of dynamical systems we have that the measure dimensions are given by the Lyapunov dimension.

(2) If (K, w_i), $i = 1, \ldots, N$, is an h.i.f.s. with attractor A such that each w_i is

a similitude with contractivity s_i and if D is the unique solution of $\sum_{i=1}^{N} s_n^D = 1$, then $H(A) \leq C(A) \leq D$ while $H(A) = C(A) = D$ in the event $w_i(A) \cap w_j(A) = \varnothing$ for $i \neq j$ [H], [BD] (see also [Mc] and [Be]). Since A is the support of the measure μ we have $H(\mu) \leq H(A)$. It is easy to verify [Mas] that $H(\mu)$ is maximized by the probability vector $(p_i^*, \ldots, p_N^*)$ where $p_i^* = s_i^D$ and $D = H(A)$. With this choice ofprobabilities $H(\mu) = H(A)$.

4. An Example with Nonsimilitudes

Here we consider an example using affine maps which are not similitudes. Let $0 < a < \frac{1}{3}$ and $a < b < \frac{1}{2}$. Define $w_i: \mathbf{R}^2 \to \mathbf{R}^2$, $i = 1, 2, 3$, by $w_i(x, y) = (ax, bx) + (c_i, d_i)$ where $(c_1, d_1) = (0, 0)$, $(c_2, d_2) = (0, 1-b)$, and $(c_3, d_3) = (1-a, \gamma)$ for some constant γ. Let S denote the unit square $[0, 1] \times [0, 1]$. Figure 1 shows the action of w_i, $i = 1, 2, 3$, on S. Let $K = [0, 1] \times [-L, L]$ with L chosen large enough so that $w_i(K) \subset K$, $i = 1, 2, 3$, and let A be the attractor for the h.i.f.s. $\{K, w_i: i = 1, 2, 3\}$. For this section we only consider the probabilities $p_1 = p_2 = p_3 = \frac{1}{3}$. Let μ be the associated invariant measure. Observe that the Lyapunov exponents for the associated dynamical system are $\lambda_1 = \log 3$, $\lambda_2 = \log b$, and $\lambda_3 = \log a$. Note that this is the set of probabilities which maximizes $\Lambda(\nu)$. In the previous section we found that for similitudes $\max(\Lambda(\nu)) + 1 = H(A) = C(A)$. However, in the case of general affine maps the situation is more complicated. For the h.i.f.s. $(K, w_i: i = 1, 2, 3)$ we find:

(a) infinitely many values of γ such that $\Lambda(\nu) - 1 > C(A)$, $H(A) \geq C(\mu)$, $H(\mu) = C(\nu) - 1$, $H(\nu) - 1$, respectively, so that $\Lambda(\nu)$ does not give any measure dimension, however,

(b) $\Lambda(\nu) = C(A) + 1$ (for the choice of probabilities $p_1 = p_2 = p_3 = \frac{1}{3}$) for Lebesgue almost every γ.

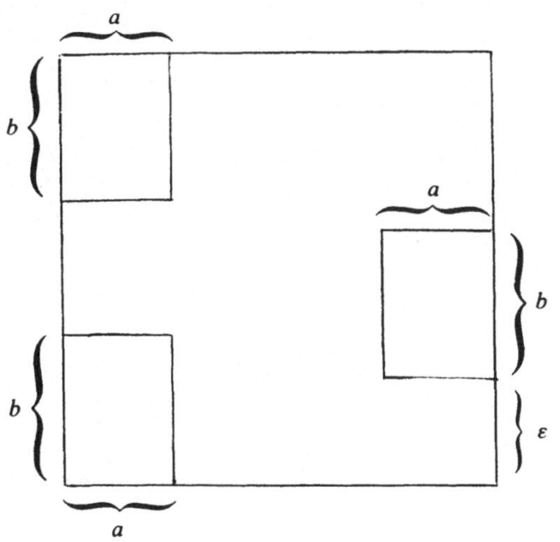

Fig. 1

Let A_p denote the y-projection of A: $A_p = \{y: (x, y) \in A \text{ for some } x\}$. Observe that A_p is the attractor for the h.i.f.s. constructed from the maps $u_1(x) = bx$, $u_2(x) = bx + (1-b)$, and $u_3(x) = bx + \gamma$. We next show that the box dimensions $C(A)$ and $C(A_p)$ are related.

Lemma 4.

$$C(A) = \frac{\log(3)}{\log(1/a)} + C(A_p)\left(1 + \frac{\log(b)}{\log(1/a)}\right).$$

Proof. Note that $C(A)$ is independent of the norm chosen to generate the ε-balls and is also independent of whether these balls are open or closed. Here we use, for convenience, ε-balls which are closed $\varepsilon \times \varepsilon$ squares with sides parallel to the axes. Let $N(\varepsilon)$ be the minimum number of such ε-balls needed to cover A and let $N_p(\varepsilon)$ be the minimum number of closed ε-intervals needed to cover A_p. The portion of A contained in the $a^k \times b^k$ rectangle $w_{i_1} \circ \cdots \circ w_{i_k}(S)$ has a minimal cover consisting of $N_p((a/b)^k)$ a^k-balls as shown in Fig. 2. Thus we have

$$N(a^k) = 3^k N_p((a/b)^k).$$

Since

$$\lim_{k\to\infty} \frac{\log(a^{k+1})}{\log(a^k)} = 1$$

we may replace $\limsup_{\varepsilon\to 0}$ by $\limsup_{k\to\infty}$ to get

$$\begin{aligned} C(A) &= \limsup_{k\to\infty} \frac{\log N(a^k)}{\log(a^{-k})} \\ &= \frac{\log 3}{\log(1/a)} + \limsup_{k\to\infty} \frac{\log N_p((a/b)^k)}{\log((a/b)^{-k})} \frac{\log(a/b)}{\log(a)} \\ &= \frac{\log 3}{\log(1/a)} + C(A_p)\left(1 + \frac{\log(b)}{\log(1/a)}\right). \end{aligned}$$

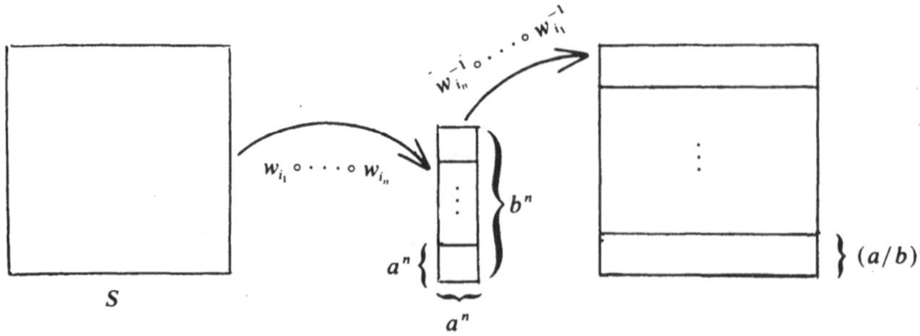

Fig. 2

If $b<\frac{1}{3}$ then the Lyapunov dimension of ν is given by

$$\Lambda(\nu)=1+\frac{\log 3}{\log(1/b)}$$

and if $\frac{1}{3}<b$

$$\Lambda(\nu)=2+\frac{\log 3+\log b}{\log(1/a)}.$$

■

First we consider the case $\gamma=0$. Here $u_1(x)=u_3(x)=bx$ and $C(A_p)$ is given by $2b^{C(A_p)}=1$ so that $C(A_p)=\log 2/\log(1/b)$. It follows from Lemma 4 that $1+C(A)<\Lambda(\nu)$. However, recall that $H(\nu)=1+H(\mu)\leq 1+C(\mu)\leq 1+C(A)<\Lambda(\nu)$ so that $C(\nu)$, $H(\nu)\neq\Lambda(\nu)$. More generally the above situation holds whenever $u_{i_1}\circ\cdots\circ u_{i_k}=u_{j_1}\circ\cdots\circ u_{j_k}$ for some $\{i_m\}_{m=1}^k\neq\{j_m\}_{m=1}^k$. This occurs for infinitely many γ.

We might expect $C(A)$ to depend continuously on γ, however this is not the case.

Theorem 3. $C(A)+1=\Lambda(\nu)$ *for Lebesgue almost every* γ.

Proof. We first find $C(\mu_p)$ where μ_p is the invariant measure associated with u_1, u_2, and u_3. We need the following result of [Ma] (see also Theorem 6.8 of [F]). Let E be a Souslin subset of $\mathbf{R}^2$ with $H(E)=s$. Let proj_θ denote the orthogonal projection of $\mathbf{R}^2$ onto L_θ, the line through the origin making an angle θ measured (for our convenience) in the clockwise direction from the y-axis. Then we have

Theorem 4. $H(\text{proj}_\theta E)=\min\{1, s\}$ *for Lebesgue almost every* $\theta\in[0,\pi)$.

To use the above result we consider the h.i.f.s. consisting of the similitudes $v_i:\mathbf{R}^2\to\mathbf{R}^2$, $i=1,2,3$, defined by $v_i(x,y)=(bx,by)+(c_i,d_i)$ with $\gamma=0$. Let B be the associated attractor. Then $H(B)=\log(3)/\log(1/b)$. The usefulness of B follows from the observation that $\text{proj}_\theta B$ and A_p are similar (i.e., related by a constant scaling factor) if $\tan\theta=\gamma$. Let $D_p=\min\{1,\log(3)/\log(1/b)\}$. Since B is a Borel set it is a Souslin set and by Theorem 4 we have $H(\text{proj}_\theta B)=D_p$ for Lebesgue almost every θ. Thus $H(A_p)=D_p$ for Lebesgue almost every γ.

However, as noted in the previous section $C(A_p)\leq D_p$ and so $H(A_p)=C(A_p)=D_p$ for almost every γ. It then follows from Lemma 4 that $\Lambda(\nu)=C(A)+1$ for Lebesgue almost every γ.

Acknowledgments. We would like to thank Michael Barnsley for encouraging discussions. We would also like to thank Peter Massopust for discussions and help with regard to the main result. Finally, we would like to thank Tim Bedford and Elles Withers for suggesting the use of Theorem 5 to calculate $H(A_p)$.

References

[B] M. F. Barnsley (1986): *Fractal functions and interpolation.* Constr. Approx., **2**:303–329.

[BD] M. F. Barnsley, S. Demko (1985): *Iterated function systems and the global construction of fractals.* Proc. Roy. Soc. London Ser. A, **399**:243–275.

[BDEG] M. F. Barnsley, S. Demko, J. Elton, J. S. Geronimo (to appear): *Invariant measures for Markov processes arising from iterated function systems with place-dependent probabilities.* Ann. Inst. H. Poincaré.

[Be] T. Bedford (1984): Crinkly Curves, Markov Partitions and Dimensions. Ph.D. Thesis, University of Warwick.

[BEHL] M. F. Barnsley, V. Ervin, D. Hardin, J. Lancaster (1986): *Solution of an inverse problem for fractals and other sets.* Proc. Nat. Acad. Sci. U.S.A., **83**:1975–1977.

[DHN] S. Demko, L. Hodges, B. Naylor (1985): *Construction of fractal objects with iterated function systems.* Comput. Graphics, **19**:271–278.

[DS] P. Diaconis, M. Shahshahani (1986): *Products of random matrices and computer image generation.* Contemp. Math., **50**:173–182.

[F] K. J. Falconer (1985): The Geometry of Fractal Sets. Cambridge: Cambridge University Press.

[FOY] J. D. Farmer, E. Ott, J. A. Yorke (1983): *The dimension of chaotic attractors.* Physica, **70**:153–180.

[H] J. Hutchinson (1987): *Fractals and self-similarity.* Indiana Univ. Math. J., **30**:731–747.

[L] F. Ledrappier (1981): *Some relations between dimension and Lyapunov exponents.* Comm. Math. Phys., **81**:229–238.

[M] B. Mandelbrot (1982): The Fractal Geometry of Nature. San Francisco: Freeman.

[Ma] J. M. Marstrand (1954): *Some fundamental geometrical properties of plane sets of fractional dimensions.* Proc. London Math. Soc. (3), **4**:257–302.

[Mas] P. Massopust (1986): Space Curves Generated by Iterated Function Systems. Thesis, Georgia Institute of Technology.

[Mc] C. McMullen (1984): *The Hausdorff dimension of general Sierpinski carpets.* Nazoyo Math. J., **96**:1–10.

[P] S. Pelikan (1984): *Invariant densities for random maps of the interval.* Trans. Amer. Math. Soc., **281**:813–815.

[Y] L. S. Young (1982): *Dimension, entropy, and Lyapunov exponents.* Ergodic Theory Dynamical Systems, **2**:109–124.

J. S. Geronimo
School of Mathematics
Georgia Institute of Technology
Atlanta
Georgia 30332
U.S.A.

D. P. Hardin
Mathematics Department
Vanderbilt University
Nashville
Tennessee
U.S.A.

Constr. Approx. (1989) 5: 99–115

CONSTRUCTIVE APPROXIMATION

Embedded Continued Fractals and Their Hausdorff Dimension

J. Harrison

Abstract. A continued fractal $Q_\alpha \subset \mathbf{R}^2$ is a curve which is associated to a real number $\alpha \in [0, 1]$. Properties of the continued fraction expansion of α appear as geometrical properties of Q_α. It is shown how number theoretic properties of α affect topological and geometric properties of Q_α such as existence, continuity, Hausdorff dimension, and embeddedness.

Introduction

A continued fractal is a curve Q_α in Euclidean space $\mathbf{R}^2$ associated to a real number $\alpha \in [0, 1]$. Properties of the continued fraction expansion of α appear as geometrical properties of Q_α. For example, we know that α is a quadratic irrational if and only if α has periodic continued fraction. Its continued fractal is self-similar [see H4]. In this paper we study how number theoretic properties of α affect topological and geometric properties of Q_α such as existence, continuity, Hausdorff dimension, and embeddedness.

Continued fractals appear as strange attractors in smooth dynamical systems. The first known example was the key element in the construction of a smooth $C^{2+\varepsilon}$ vector field on the three sphere S^3 with no zeros and no closed integral curves (see [H1] and [H3]). Continued fractals can be Julia sets for *diffeomorphisms* f of the two-sphere S^2. They may be contrasted with the original fractal Julia sets of Julia, Fatou, Sullivan *et al.*, which arise from *noninvertible* mappings (see [B], for example). The invertibility of f gives it a basic position in the world of two-dimensional dynamics.

1. The Embedding $h: I \to \mathbf{R}^2$

We give two descriptions of h. The first is geometric. It provides a guide for programming computers to draw $h(I)$. The second is analytic and is used for proving the theorems.

Date received: September 11, 1987. Communicated by Michael F. Barnsley.
AMS classification: 11A55, 28A05.
Key words and phrases: Continued fractions, Diophantine type, Fractals, Hausdorff dimension.

Notation

Let $\mathbf{R}$ denote the real numbers and $Q \subset \mathbf{R}$ the rationals. Let π_1 and π_2 denote the projections onto the first and second factors of $\mathbf{R}^2$, respectively. If $x \in \mathbf{R}$, define $\langle x \rangle = x \pmod 1$, $\text{int}(x) = x - \langle x \rangle$, and $\|x\| =$ distance to the nearest integer. Let $I = [0, 1)$. Any function $h: I \to \mathbf{R}^2$ canonically induces a function $\tilde{h}: S^1 \to S^1 \times \mathbf{R}^1$. (Define $h^*: \mathbf{R}^1 \to \mathbf{R}^2$ by $h^*(x) = h(\langle x \rangle) + (\text{int}(x), 0)$. This h^* induces a function $\tilde{h}: S^1 \to S^1 \times \mathbf{R}^1$ via the identifications $x \sim x + n$ on $\mathbf{R}^1$ and $(x, y) \sim (x + n, y)$ on $\mathbf{R}^2$.) For simplicity of notation, we work entirely with $h: I \to \mathbf{R}^2$. Fix $\alpha \in \mathbf{R}$ and define

$$x_n = \langle n\alpha \rangle, \qquad n \geq 0.$$

Normalizing Constants. Let $g: Z^+ \to \mathbf{R}$ be a monotone decreasing function. For $k \geq 1$, we define two normalizing constants:

$$g(0) = \sum (-1)^n g(n), \qquad n = 1, \ldots, k$$

and

$$m_k = 1 + \sum g(n), \qquad n = 0, \ldots, k.$$

The Computer Algorithm

The algorithm depends only on the sequence x_n and weights $g(n)$. Let $x_{n_0} = 0$, $x_{n_1}, \ldots, x_{n_k}$ denote the first $k+1$ terms of the sequence x_n, ordered from left to right in I. Let

$$I_{n_i} = (x_{n_i}, x_{n_{i+1}}), \qquad i = 0 \text{ to } k-1.$$

In the plane, draw a horizontal line segment with left endpoint the origin $(0, 0)$ and length $m_k|I_{n_0}|$. Attach to its right endpoint a line segment sloping backward with fixed, acute angle $\pi/4$ and length $g(n_1)/(\cos \pi/4) = \sqrt{2}g(n_1)$. We call this segment a *diagonal*, Δ_{k,n_1}. It points down if n_1 is even and up if n_1 is odd. Attach to the free endpoint of Δ_{k,n_1}, forming the angle $\pi/4$, a horizontal line segment of length $m_k|I_{n_1}|$. Attach to the free endpoint of this second horizontal line a diagonal Δ_{k,n_2} of length $\sqrt{2}g(n_2)$ again sloping backward with angle $\pi/4$ and up or down according to whether n_2 is odd or even. Continue until k horizontal and diagonal lines have been drawn, one pair for each of the intervals I_{n_i}. Finally, draw a diagonal $\Delta_{k,0}$ with endpoint the rightmost endpoint of the curve and length $\sqrt{2}g(0)$. Call the resulting curve Q_k (see Figs. 1-3).

We prove that for some α and g, there is a subsequence k_i such that the curves Q_{k_i} converge to a limit curve Q. In this case we denote

$$\Delta_n = \lim_{i \to \infty} \Delta_{k_i,n}.$$

An Analytical Definition of Q_k

Let $\rho: I \to I$ be a monotonic, continuous mapping such that $\rho^{-1}(x_n)$ is an interval $\Delta'_n = [y'_n, z'_n]$, $n \geq 0$, and ρ is $1-1$ on the complement of the $\{\Delta'_n\}$. (If x_n is dense,

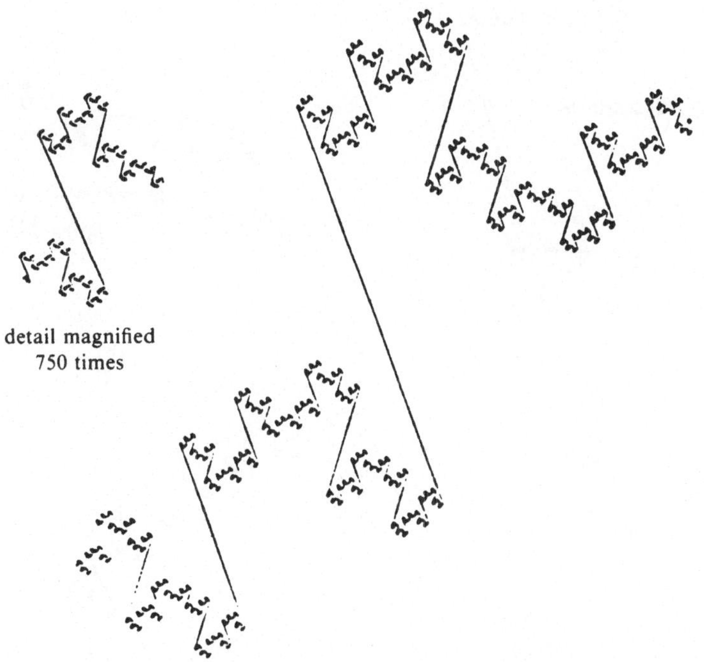

Fig. 1. Continued fractal for $\alpha = (\sqrt{5}-1)/2 =$ the Golden Mean, $g(n) = 0.7/n^{0.7}$, $n \le 5000$.

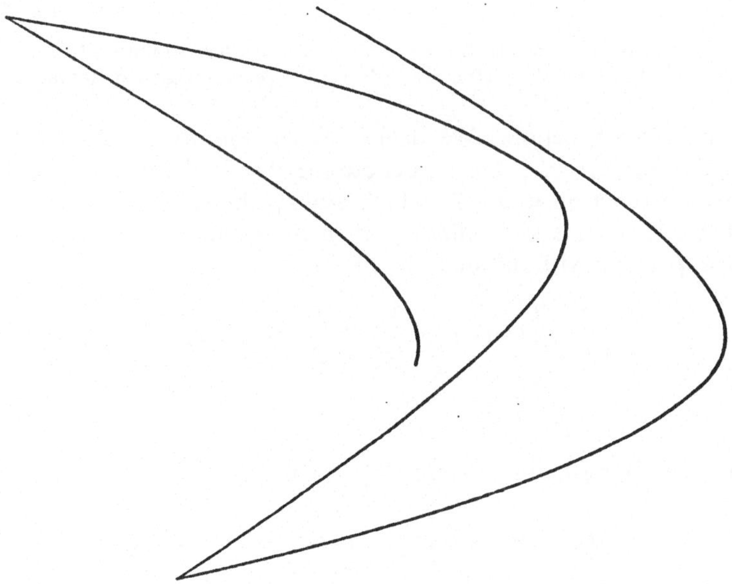

Fig. 2. Continued fractal for $x_n = n\alpha$, $\alpha = \pi/2$ (detail magnified 75 times), $g(n) = 0.7/n^{0.7}$, $n \le 5000$.

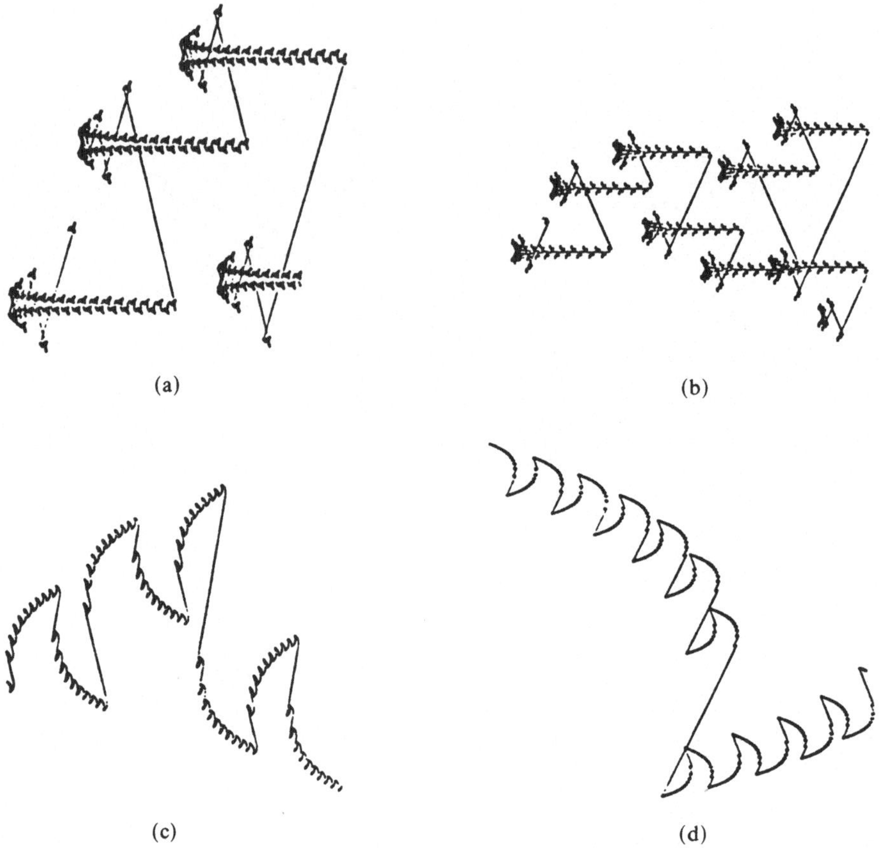

Fig. 3. Details of continued fractals: (a) $\alpha = 2^{1/4}$ (detail magnified 10 times), (b) $\alpha = 2^{1/4}$ (detail magnified 2000 times), (c) $\alpha = 7^{1/4}$ (full curve), (d) $\alpha = 7^{1/4}$ (detail magnified 10 times).

then ρ is a Cantor function.) We define immersions $h_k\colon I \to \mathbf{R}^2$, $k \geq 1$, so that $h_k(I) = Q_k$ and $h_k(\Delta'_n) = \Delta_n$. Let χ_A denote the standard characteristic function. We also need a function $d\colon I \to \mathbf{R}^2$ which adjusts the location of $h_k(x)$ for x in a segment Δ'_n. It plays a very minor part in the estimates. (Indeed, the reader may assume $d(x) = 0$ without much loss.)

$$d(x) = \begin{cases} (-1)^n g(n) \dfrac{|x - y'_n|}{|z'_n - y'_n|} & \text{if } x \in \Delta'_n, \\ 0 & \text{if } x \notin \bigcup \Delta'_n. \end{cases}$$

Definition. Let $W = [0, \rho(x))$. Define $h_k\colon I \to \mathbf{R}^2$ by

$$\pi_1 h_k(x) = m_k |W| - \sum_{n=0}^{k} \chi_W \langle i\alpha \rangle g(n) - |d(x)|,$$

$$\pi_2 h_k(x) = \sum_{2n+1 \leq k} \chi_W \langle (2n+1)\alpha \rangle g(2n+1) - \sum_{2n \leq k} \chi_W \langle 2n\alpha \rangle g(2n) - d(x).$$

Define

$$h = \lim h_k \qquad \text{as} \quad k \to \infty.$$

It is not hard to see that if $\sum g(n) < \infty$, then h exists and is continuous. If $\sum g(n) = \infty$, then the construction poses more difficult and interesting problems. The critical exponent of the series $\sum g(n)$ is the Hausdorff dimension in the embedded examples. Hence, the "longer" the curve, the higher the Hausdorff dimension and the more "unfolded" the sequence x_n becomes.

Remark. Q_k may be defined for *any* 1-1 sequence x_n and weight function $g(n)$. In this paper we restrict ourselves to $x_n = \langle n\alpha \rangle$, mod 1.

2. Number Theory Prerequisites

Definition 2.1. Let a_n be a sequence of integers ≥ 1 and

$$\frac{p_n}{q_n} = \left[\cfrac{1}{a_2 + \cfrac{1}{a_2 + \cfrac{1}{a_3 + \cdots \cfrac{1}{a_n}}}} \right],$$

where $(p_n, q_n) = 1$. Then

$$\alpha = \lim_{n \to \infty} \frac{p_n}{q_n}$$

exists and is a positive irrational <1. We write $\alpha = (a_1, a_2, a_3, \ldots)$. The fractions p_n/q_n are *rational convergents of* α. The integers p_n and q_n satisfy the recursive relations:

$$p_0 = 0, \quad p_1 = 1, \quad \text{and} \quad p_n = a_n p_{n-1} + p_{n-2};$$

$$q_0 = 1, \quad q_1 = a_1, \quad \text{and} \quad q_n = a_n q_{n-1} + q_{n-2}.$$

Define $r_0 = 1$, $r_1 = \alpha$, and $r_{n-1} = a_n r_n + r_{n+1}$.

The first lemma follows readily from these definitions.

Lemma 2.2.

(i) $q_n r_n + q_{n-1} r_{n+1} = 1$.
(ii) $q_n p_{n-1} - p_n q_{n-1} = (-1)^n$.
(iii) $p_n - q_n \alpha = (-1)^{n+1} r_{n+1}$.
(iv) *If* $\alpha = [c, c, c, \ldots]$ *then* $r_n = \alpha^n$.

Lemma 2.3. *If $\alpha \in \mathbf{R}\backslash Q$ has continued fraction expansion $[a_1, a_2, \ldots]$, then for each $n \geq 1$*

$$\left|\alpha - \frac{p_n}{q_n}\right| = \frac{1}{a_n^2(a_{n+1} + (a_{n+2}, a_{n+3}, \ldots) + (a_n, \ldots, a_1))},$$

where p_n/q_n is the n*th convergent of α.*

Proof. See Lemma 1.5 of [HN]. ■

Let $n \in \mathbf{Z}^+$. Define

$$t_n = q_n + q_{n-1} - 1.$$

We define W_n to be the collection of intervals in I complementary to $\{\langle 0\alpha\rangle, \langle 1\alpha\rangle, \ldots, \langle t_n\alpha\rangle\}$: each of these intervals is a *W_n-interval* and $\langle j\alpha\rangle$, $0 \leq j \leq t_n$, is a *W_n-point.* By Lemma 2.2(iii) $\langle q_{2n+1}\alpha\rangle$ converges monotonically to 0 and $\langle q_{2n}\alpha\rangle$ converges monotonically to 1. Let $I_0(n)$ be the W_n-interval with endpoints $\langle q_{n-1}\alpha\rangle$ and an endpoint of I and $J_0(n)$ the W_n-interval bounded by $\langle q_n\alpha\rangle$ and the other endpoint of I. Define

$$R_\alpha(x) = x + \alpha \pmod 1.$$

Lemma 2.4. *The collection W_n consists of the first q_n iterates of $I_0(n)$ and the first q_{n-1} iterates of $J_0(n)$ under the transformation R_α. In particular, all W_n-intervals have length r_n or r_{n+1}.*

Proof. This follows from Lemma 2.2(i). (See also Lemma 1.5 of [H2].) ■

Definition. An irrational number α satisfies a *Diophantine condition σ* if there exists $c > 0$ such that

$$\left|\alpha - \frac{p}{q}\right| > \frac{c}{q^{1+\sigma}} \qquad \text{for every } \frac{p}{q} \in Q.$$

α has *Diophantine type δ if*

$$\delta = \underline{\lim}\{\sigma\colon \alpha \text{ satisfies a Diophantine condition } \sigma\}.$$

Estimates on the Discrepancy of $\langle n\alpha\rangle$

The remainder of this section is devoted to estimates of "weighted discrepancy" which are fundamental to the study of continued fractals.

Theorem 2.5

(i) *Let $\alpha \in \mathbf{R}\backslash Q$ and p/q be a rational convergent of α. If J is an interval of I with $|J| = \|q\alpha\|$, then for every $0 \leq s < t$*

$$\left|\sum_{i=s}^{t-1} (\chi_J\langle i\alpha\rangle - |J|)\right| < 2.$$

(ii) *Assume $\alpha \in \mathbf{R}\backslash Q$ has Diophantine type δ and $\varepsilon > 0$. There exists $C > 0$ such that if U is an interval of I and $0 \le s < t$, then*

$$\left|\sum_{i=s}^{t-1} (\chi_U\langle i\alpha\rangle - |U|)\right| < C(t-s)^{1-(1/\delta)+\varepsilon}.$$

Proof. For (i), see [K] or Theorem 2.2(i) of [H2]. (ii) follows immediately from Theorem 3.2 of [KN, p. 123]. ∎

The next lemma is used in weighted versions of Theorem 2.5.

Lemma 2.6. *Let $C > 0$ and $f(i)$ be a monotone increasing sequence of positive real numbers. Let $b(i)$ be a sequence of real numbers satisfying*

$$\left|\sum_{i=j}^{j+N-1} b(i)\right| < Cf(N) \qquad \textit{for all} \quad j, N \ge 0.$$

Let $g(i) > 0$ be a monotone decreasing sequence. Then:

(i)
$$\left|\sum_{i=s}^{t-1} b(i)g(i)\right| \le Cf(t-s)g(s) \qquad \textit{for all} \quad 0 \le s \le t.$$

(ii) *Define n and r by $2^n \le t-1 < 2^{n+1}$ and $2^r \le s < 2^{r+1}$. Then*

$$\left|\sum_{i=s}^{t-1} b(i)g(i)\right| \le C\left|\sum_{\lambda=r}^{n} f(2^\lambda)g(2^\lambda)\right|.$$

Proof. This is essentially "summation by parts." (For details, see Proposition 2.3 of [H2].) ∎

Corollary 2.7.

(i) *Let $\alpha \in \mathbf{R}\backslash Q$. If $J \subset I$ is an interval with $|J| = \|q_n\alpha\|$ and $g\colon \mathbf{Z} \to \mathbf{R}^+$ is monotone decreasing, then*

$$\left|\sum_{i=1}^{t-1} (\chi_J\langle i\alpha\rangle - |J|)g(i)\right| < 2g(s).$$

(ii) *Assume $\alpha \in \mathbf{R}\backslash Q$ has Diophantine type δ. Let $\varepsilon > 0$ and $\eta = 1-(1/\delta)+\varepsilon$. There exists $C > 0$ such that if U is an arbitrary interval of I and $g(i) \le 1/i^\gamma$ where $\gamma > \eta + 1/2\delta$, then*

$$\left|\sum_{i=s}^{t-1} (X_U\langle i\alpha\rangle - |U|)g(i)\right| < Cs^{\eta-\gamma}.$$

Proof. (i) This follows immediately from Kesten's theorem (Theorem 2.5(i)) and summation by parts (Theorem 2.6(i)).

(ii) Let C_1 be the constant depending on α and ε obtained from Theorem 2.5(ii). Apply Lemma 2.6(ii) to $b(i)=\chi_U\langle i\alpha\rangle-|U|$, $g(i)\le 1/i^\gamma$, and $f(i)=i^\eta$ to obtain

$$\left|\sum_{i=s}^{t-1} b(i)g(i)\right| < C_1\left|\sum_{\lambda=r}^{n} 2^{\lambda(\eta-\gamma)}\right| < C_1C_2 s^{\eta-\gamma}.$$

Note that C_2 depends on δ since $1/2\delta<\lambda-\eta$. Let $C=C_1C_2$. Then C depends on α and ε. ■

3. Sufficient Conditions for Q to Exist

Most irrational numbers are Diophantine. If α has Diophantine type δ we show that the curve Q generated by a monotone function $g(n)\le c/n^\gamma$ and α exists and is continuous for $\gamma>1-1/2\delta$. The Hausdorff dimension of Q is bounded above by $1/\gamma$ if $\frac{1}{2}<\gamma<1$ and by 1 if $\lambda\ge 1$. This bound is sharp since there are examples with Hausdorff dimension $1/\gamma$ for $\frac{1}{2}<\gamma\le 1$.

We prove that the subsequence of Q_{t_k} converges to a limit curve where $t_k=q_k+q_{k-1}-1$. Henceforth, for simplicity of notation, set

$$Q_k=Q_{t_k},$$
$$h_k=h_{t_k},$$
$$m_k=m_{t_k}.$$

Theorem 3.1. *Let $\alpha\in\mathbf{R}\backslash Q$ have Diophantine type δ. Assume $\gamma>1-1/2\delta$. Let Q_k be the sequence of curves generated by $g(n)\le c/n^\gamma$ and α. Then $Q=\lim Q_{t_k}$ exists and is continuous.*

Proof. Recall that $Q_k=h_k(I)$. Let U be an interval in I with endpoints $p<q$. Define

$$H_k=H_k(U)=\pi_1h_k(q)-\pi_1h_k(p),$$
$$V_k=V_k(U)=\pi_2h_k(q)-\pi_2h_k(p).$$

See Fig. 4.

In order to prove that h_k converges uniformly to a continuous function h, it suffices to show that H_k and V_k converge uniformly over intervals of I. Let C be the constant of Theorem 2.7(ii). Recall

$$\Delta'_n=(y'_n, z'_n)=\rho^{-1}\langle n\alpha\rangle.$$

In order to prove Theorem 3.1 we verify the uniform convergence of H_k and V_k over two basic types of intervals U with endpoints $p<q$:

(3.1) $$p\notin\bigcup[y'_n, z'_n) \quad\text{and}\quad q\notin\bigcup(y'_n, z'_n];$$

(3.2) $$[p,q]\subset[y'_n, z'_n] \quad \text{for some} \quad n\in\mathbf{Z}.$$

Uniform convergence over all intervals of I follows.

In the hypotheses we have $\gamma>1-1/2\delta$, $\delta\ge 1$. Choose $\varepsilon>0$ with

$$\gamma>1+\varepsilon-1/2\delta.$$

Define $\eta=1+\varepsilon-1/\delta$.

Preliminary Estimate. Let $c=1$. Recall the normalizing constant

$$m_k = 1+\sum g(n)+(-1)^n g(n), \qquad n=1,\ldots,t_k.$$

Then

$$m_k < 1+\frac{t_k^{1-\gamma}}{1-\gamma} < 1+\frac{2q_k^{1-\gamma}}{1-\gamma}.$$

Since $r_k q_k < 1$ (by Lemma 2.2) it follows that $m_k r_k \to 0$ as $k\to\infty$. Therefore, for $\beta>0$, there exists N such that if $t_k \geq N$, then

$$\max\{m_k r_k, 2Ct_k^{\eta-\gamma}, 2t_k^{-\gamma}\} \leq \frac{\beta}{3}.$$

Proof of the Uniform Convergence of H_k. Let U be an interval of type (3.1) with endpoints $p<q$. Then $d(q)=0$. Let $W=\operatorname{int}(\rho(U))$ and $N<t_j<t_k$. It follows that

$$\left(H_k(U)=1+\sum_{n=0}^{t_k} g(n)\right)|W| - \sum_{n=m}^{t_k} \chi_W\langle n\alpha\rangle g(n), \tag{3.3}$$

where $m=\min\{n: \langle n\alpha\rangle \in \operatorname{int}(W)\}$. (Whenever $m>t_k$, the last sum is zero.) Assume $m\leq t_j<t_k$. Apply Theorem 2.7(ii). Then

$$|H_j-H_k| = \left|\sum_{n=t_j}^{t_k} (|W|-\chi_W\langle n\alpha\rangle)g(n)\right| < Ct_j^{\eta-\gamma} < \beta.$$

Assume $t_j<t_k<m$. Then $|W|\leq t_k$. Since the last sum in (3.3) is 0 for both H_j and H_k we have

$$|H_j-H_k| = |m_j-m_k||W| < m_k t_k < \beta.$$

If $t_j<m\leq t_k$, let l satisfy $t_j\leq t_l<m\leq t_{l+1}\leq t_k$. Then $|W|\leq r_l$. The last sum in (3.3) for H_j is zero so

$$\begin{aligned}
|H_j-H_k| &= \left||W|\sum_{n=t_j+1}^{t_k} g(n)+\sum_{n=t_l+1}^{t_k} \chi_W\langle n\alpha\rangle g(n)\right| \\
&= \left||W|\sum_{n=t_j+1}^{t_l} g(n)+\sum_{n=t_l+1}^{t_k} (|W|-\chi_W\langle n\alpha\rangle)g(n)\right| \\
&< m_l r_l + Ct_l^{\eta-\gamma} \qquad \text{(by Theorem 2.7(ii))} \\
&< \beta.
\end{aligned}$$

Thus H_k converges uniformly over the intervals U of type (3.1).

Now assume $U\subset\Delta_n'$. In general, if $n\leq t_l$ then $|H_l|\leq g(n)$. If $n>t_l$, then $|H_l|\leq m_l r_l$. Therefore, if $t_j<t_k<n$,

$$\begin{aligned}
|H_j-H_k| &< \max\{|H_j|, |H_k|\}, \\
&\leq \max\{m_j r_j, m_k r_k\} \\
&< \beta.
\end{aligned}$$

If $n \le t_j < t_k$, then $|H_j - H_k| = 0$. Finally, if $t_j < n \le t_k$,

$$\begin{aligned}|H_j - H_k| &< \max\{|H_j|, |H_k|\}\\ &\le \max\{m_j r_j, g(n)\}\\ &< \max\{m_j r_j, g(t_j)\}\\ &< \beta.\end{aligned}$$

Proof of Uniform Conference of V_n. Let $U \subset I$ be an interval of type (3.1) and $W = \text{int}(\rho(U))$. Then

$$V_k = \sum_{2n+1 \le t_k} \chi_W\langle(2n+1)\alpha\rangle g(2n+1) - \sum_{2n \le t_k} \chi_W\langle 2n\alpha\rangle g(2n). \tag{3.4}$$

Let $N < t_j < t_k$. Then

$$\begin{aligned}|V_j - V_k| &= \left|\sum_{t_j < 2n+1 \le t_k} \chi_W\langle(2n+1)\alpha\rangle g(2n+1) - \sum_{t_j < 2n \le t_k} \chi_W\langle 2n\alpha\rangle g(2n)\right|\\ &= \left|\sum_{t_j < 2n+1 \le t_k} (\chi_W\langle(2n+1)\alpha\rangle - |W|) g(2n+1)\right.\\ &\quad + \sum_{t_j < 2n \le t_k} (|W| - \chi_W\langle 2n\alpha\rangle) g(2n)\\ &\quad \left. + |W| \sum_{t_j < 2n, 2n+1 \le t_k} (g(2n+1) - g(2n))\right|.\end{aligned}$$

Apply the triangle inequality and Theorem 2.7(ii) to estimate the first two sums. (The third sum is bounded by $2g(j)$ since $|W| < 1$.) Thus

$$|V_j - V_k| < 2Ct_j^{\eta-\gamma} + 2g(t_j) < \beta.$$

Suppose $U \subset \Delta_n'$. Then V_l is zero for $t_l < n$. If $n \le t_l$ then V_l is a constant depending only on U. It is bounded by $g(n)$. Therefore, if $n \le t_j < t_k$ or $t_j < t_k < n$, then $|V_j - V_k| = 0$. If $t_j < n \le t_k$, then $|V_k - V_j| = |V_k| < g(n) < g(j) < \beta$. Hence V_k converges uniformly.

The proof for arbitrary $c > 0$ is similar. ∎

4. Embedded Continued Fractals and Number Theory

It is more delicate to show Q is embedded. In [H2] there is a proof for $\alpha = \sqrt{2} - 1$. Here, we give a new proof for a class of numbers containing α. The definition of g given below makes it possible to provide sharp estimates for the embedding.

Definition of g. Let $\alpha \in \mathbf{R} \backslash Q$, $c > 0$, and $\frac{1}{2} < \gamma < 1$. Define

$$g(n) = c\alpha^{k\gamma} \qquad \text{for} \quad q_{k-1} \le n < q_k.$$

Theorem 4.1. *Assume $\alpha = [N, N, N, \ldots]$, N even. There exists constants $c_N > 0$ and $\gamma_N < 1$ such that if $0 < c < c_N$ and $\gamma_N < \gamma < 1$, then $h = h_{\gamma,c}\colon I \to \mathbf{R}^2$ is an embedding onto Q.*

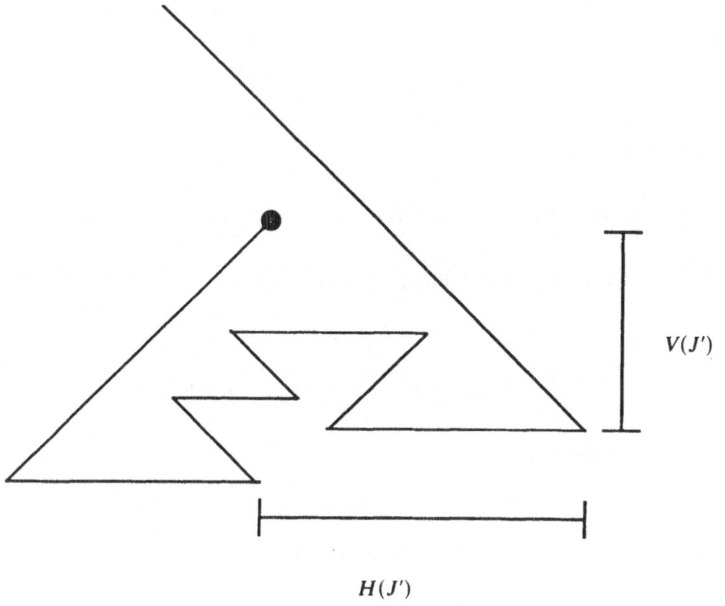

Fig. 4

Proof. The idea of the proof is fairly simple. The line segment I is covered by W_k-intervals. Let J be one of these. We may assume $|J| = \alpha^k$, otherwise it appears as a W_{k+1}-interval. See Lemmas 2.4 and 2.2(iv). Assume that J includes the endpoint with larger index, n, say. Then $q_{k-1} \leq n < q_k$. Let $J' = \rho^{-1}(J)$. Define $H = H(J')$ and $V = V(J')$ as in the previous section. If $H > V$ then the two diagonals attached to the endpoints of $h(J')$ are disjoint. By induction the curve is embedded (see Fig. 4). ■

Let W denote the covering of I by W_k-intervals, $k \geq 1$.

Theorem 4.2.

(i) *There exist* $c_N > 0$ *and* $\frac{1}{2} < \gamma_N < 1$ *so that* $|H/V| > 1$ *for all* $0 < c < c_N$, $\gamma_N < \gamma < 1$ *and* $J \in W$.

(ii) $|H/V|$ *is uniformly bounded: given* c *and* γ, *there exists* $L > 0$ *with* $|H/V| < L$, *for all* $J \in W$.

Remark. We only need Theorem 4.2(i) for this proof; Theorem 4.2(ii) is needed later in the estimates for the Hausdorff dimension of Q.

Proof of Theorem 4.2. It follows from Theorems 3.1 and 3.3 that $H = \alpha^k + A + B$ where

$$A = \alpha^k \sum_{i=1}^{q_{k-1}-1} g(i)$$

and

$$B=\sum_{i=q_{k-1}}^{\infty}(|J|-\chi_J(\langle i\alpha\rangle))g(i).$$

By the choice of g,

$$A=c\alpha^k[(q_1-q_0)\alpha^{\gamma}+(q_2-q_1)\alpha^{2\gamma}+\cdots+(q_{k-1}-q_{k-2})\alpha^{(k-1)\gamma}].$$

To compute A, we need some simple, preliminary facts: by Lemmas 2.2(iii) and 2.3,

$$\alpha^{n+1}=|q_n\alpha-p_n|=\frac{1}{q_n(N+\alpha+p_n/q_n)}=\frac{1}{q_n(N+2\alpha+\varepsilon_n)},$$

where $|\varepsilon_n|<2/q_n^2$. Therefore

$$q_n=\frac{\alpha^{-(n+1)}}{(N+2\alpha+\varepsilon_n)},$$

since $\alpha=[N, N, N, \ldots]$, $\alpha^{-1}=N+\alpha$. Hence

$$q_n-q_{n-1}=\frac{\alpha^{-n}(N+\alpha-1)(1+\beta_n)}{(N+2\alpha)},$$

where $\beta_n\to 0$. Let $\Delta=c\alpha^{k\gamma}$. It is straightforward to verify that as $k\to\infty$

$$\frac{A}{\Delta}\to\frac{N+\alpha-1}{(N+2\alpha)(\alpha^{\gamma-1}-1)}. \tag{4.1}$$

We next estimate B; let $b(i)=|J|-\chi_J(\langle i\alpha\rangle)$. We apply summation by parts:

$$\begin{aligned}B=\lim[b(q_{k-1})+\cdots+b(q_m)]g(q_m)+b(q_{k-1})[g(q_{k-1})-g(q_{k-1}+1)]\\+[b(q_{k-1})+b(q_{k-1}+1)][g(q_{k-1}+1)-g(q_{k-1}+2)]\\+\cdots+[b(q_{k-1})+\cdots+b(q_m-1)][g(q_m-1)-g(q_m)].\end{aligned}$$

Now $\sum b(i)$ is bounded (according to Kesten's Theorem 2.5(i)) and $g(q_m)\to 0$, so the first term vanishes in the limit. Note that $g(i)-g(i-1)=0$ unless $i=q_j$. Also, $g(q_m-1)=g(q_{m-1})$. Hence,

$$\begin{aligned}B=[b(q_{k-1})+\cdots+b(q_k-1)][g(q_{k-1})-g(q_k)]\\+[b(q_{k-1})+\cdots+b(q_{k+1}-1)][g(q_k)-g(q_{k+1})]+\cdots.\end{aligned}$$

Next we substitute the identity $g(q_{n-1})-g(q_n)=\alpha^{n\gamma}(1-\alpha^{\gamma})$. Assume that k is even. According to Theorem 4.2 of [HN], we know

$$\sum_{i=q_{k-1}}^{q_n-1} b(i)=\sum_{i=0}^{q_n-1} b(i)-\sum_{i=0}^{q_{k-1}-1} b(i)=q_{k-1}(\pm\alpha^{n+1}-\alpha^k),$$

+ if n is even, − if n is odd. If k is odd, we get

$$q_{k-1}(\pm\alpha^{n+1}-\alpha^k)=q_{k-1}(\pm\alpha^{n+1}-\alpha^k),$$

− if n is even, + if n is odd. We write these in terms of α, using the above identities and get

$$\sum b(i)=\frac{(-1\pm\alpha^{n+1-k})}{(N+2\alpha+\varepsilon_k)},$$

+ if and only if k and n have the same sign. Putting all of this together yields

$$\begin{aligned} B&=\frac{\alpha^\gamma-1}{N+2\alpha+\varepsilon_k}[(-\alpha+1)\alpha^{\gamma k}+(-\alpha^3+1)\alpha^{\gamma(k+2)}+\cdots+(\alpha^2+1)\alpha^{\gamma(k+1)}\\ &\qquad+(\alpha^4+1)\alpha^{\gamma(k+3)}+\cdots]\\ &=\left[\frac{\alpha^\gamma-1}{N+2\alpha+\varepsilon_k}\right]\left[\frac{-\alpha^{1+\gamma k}+\alpha^{2+\gamma(k+1)}}{1-\alpha^{2+2\gamma}}+\frac{\alpha^{\gamma k}}{1-\alpha^\gamma}\right].\end{aligned}$$

Thus

$$\frac{B}{\Delta}\to\frac{\alpha-\alpha^\gamma(\alpha^2+\alpha)+2\alpha^{2+2\gamma}-1}{(N+2\alpha)(1-\alpha^{2+2\gamma})}. \tag{4.2}$$

Last of all, we estimate V/Δ. Recall (3.4). We have

$$V=\sum_{n=q_{k-1}}^{\infty}[(\chi_J\langle n\alpha\rangle, n\text{ odd})-(\chi_J\langle n\alpha\rangle, n\text{ even})]g(n).$$

Since N is even, the sequence q_n alternates parity, starting with $q_0=1$, $q_1=2$, $q_{n+1}=2q_n+q_{n-1}$. Therefore, if $k-1$ is odd, q_{k-1} is even. Thus the diagonals attached to J are parallel. This will pose no difficulty for the embedding. (See Fig. 5.) We study the case with $k-1$ even:

$$|V/c|=\alpha^{k\gamma}-2[\alpha^{(k+1)\gamma}+\alpha^{(k+3)\gamma}+\cdots]=\alpha^{k\gamma}-2\alpha^{(k+1)\gamma}/(1-\alpha^{2\gamma}).$$

Hence $|V/\Delta|\to 1-2\alpha^\gamma/(1-\alpha^{2\gamma})$.

Let γ_N be the solution for the limit equation $H/V=1$. As $N\to\infty$, $\gamma_N\to 1$. If $\gamma_N<\gamma<1$, there is k_N such that if $k>k_N$, then $|H/V|>1$. The choice of the constant c affects H for k small. Recall

$$H/\Delta=(\alpha^k+A+B)/\Delta=\alpha^{k(1-\gamma)}/c+(A+B)/\Delta>\alpha^{k(1-\gamma)}/c.$$

We know that $|V/\Delta|<C$ for some $C>0$ and all $k>0$. It suffices to find c so that

$$\alpha^{k(1-\gamma)}/c>C \qquad \text{for} \quad k\le k_N.$$

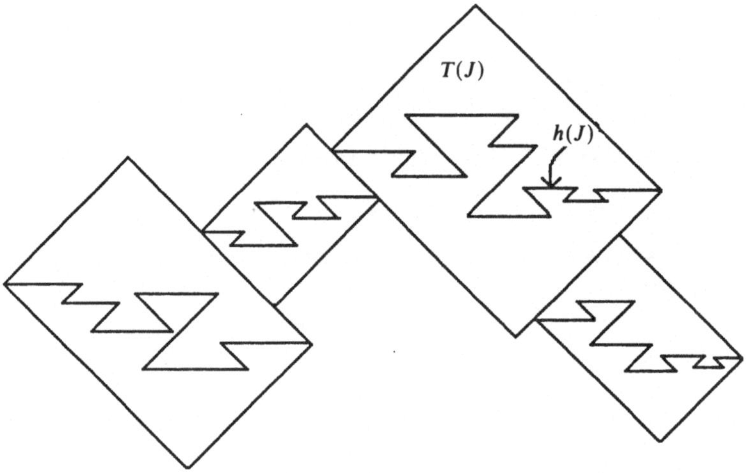

Fig. 5

It will follow that $|H/V|>1$ for all k and that h is an embedding. Fix $c_N = \alpha^k N^{(1-\gamma)}/C$. So $\alpha^{k(1-\gamma)}/c_N > C$ for $k \le k_N$. Then $|H/V|>1$ for $\gamma_N < \gamma < 1$ and $0 < c < c_N$. It is easy to see that $|H/\Delta|$ is uniformly bounded for fixed c and γ. Hence $|H/V|$ is uniformly bounded.

This completes Theorem 4.2. ■

Remarks. The constant γ_N is sharp. That is, if $\gamma < \gamma_N$, then $H/\Delta > 1$ for k sufficiently large. It follows that Q is not embedded for such γ.

Proof of Embedding. Begin with γ_N and c_N of Theorem 4.2. Let J be a W_k-interval and $H = H(J)$, $V = V(J)$, and $\Delta = \Delta(J)$ be defined as above. Recall that J contains the endpoint with the largest index. We have $|H/V|>1$. Let $T = T(J)$ be the parallelogram determined by the endpoints of $h(\text{int}(J))$, see Fig. 5. By Theorem 4.2, the smaller diagonal attached to the end of $h(\text{int}(J))$ is contained in T. This is the fundamental geometric observation for it implies that $h(J) \subset T$.

Let x and $y \in I$. Let k be the first integer such that $\rho(x)$ and $\rho(y)$ are separated by a W_k-point. Then $\rho(x) \in J_1$ and $\rho(y) \in J_2$ where J_1 and J_2 are W_k-intervals. The chain of W_k-parallelograms T are disjoint by Theorem 4.2. Therefore $h(x) \neq h(y)$.

5. Hausdorff Dimension of Q

Let Q be the curve generated by $\alpha = [N, N, N, \ldots]$, and let c and γ be as in the previous section.

Definition. Given a metric space X, for each nonnegative real number s there is a corresponding s-dimensional Hausdorff measure μ_s defined as follows. Let $B \subset X$ be an arbitrary set. The zero-dimensional measure $\mu_0 B$ is the number of points in B. For $s > 0$, $\alpha > 0$, let

$$\mu_{s,\alpha} B = \inf \sum_i [\text{diam}(B_i)]^s,$$

where the infimum is taken over all covers $\{B_i\}$ of B such that $\text{diam}(B_i) < \alpha$ for each i. Then

$$\mu_s B = \lim_{q \to 0^+} \mu_{s,\alpha} B.$$

A set B has *Hausdorff dimension* s iff $\mu_r B = 0$ for all $r > s$ and $\mu_r B = \infty$ for all $r < s$. If $\dim(B) = s$, then $\mu_s B$ is the *Hausdorff measure of B within its dimension.*

Let Γ denote the Cantor set $Q \setminus \bigcup \{\text{int}\, \Delta_n\}$.

Theorem 5.1. *The Hausdorff dimension of* Γ *is* $1/\gamma$. *The Hausdorff measure of* Γ *within its dimension is a positive, real number.*

Proof. Let $I_0 = I \setminus \{\langle n\alpha \rangle\}$. Define $g\colon I_0 \to \mathbf{R}^2$ by $g = h \circ \rho^{-1}$.

Claim. There exist constants A_1, A_2, A_3, and K_1 such that:

(1) If J is a W_k-interval, $k \geq 0$, then

$$A_2 |J|^\gamma < |g(J)| < A_1 |J|^\gamma.$$

(2) If J is an arbitrary interval of S^1, then

$$|g(J)| > A_3 |J|^\gamma.$$

(3) If $p, q \in Q$, the arc connecting p and q is contained in a disk of radius $K_1 d(p, q)$. Hence if $B \subset \mathbf{R}^2$ is a disk with $B \cap g(I_0) \neq \emptyset$, then

$$\text{cl}(g^{-1}(K_1 B)) \supset \text{interval} \supset g^{-1}(B).$$

Proof of Claim. (1) By (4.1) and (4.4), $H \to C\Delta$ for some $C > 0$. The definitions imply $\Delta = c\alpha^{k\gamma} = c|J|^\gamma$. Finally, Theorem 4.2 implies $|H/V|$ is bounded away from 0 and ∞. This establishes (1).

(2) Let I be a W_k-interval contained in J, where k is minimal. Then

$$(2N+1)|I| > |J| > |I|$$

since any $2N+1$ adjacent W_k-interval contains a W_{k-1}-interval. Then

$$|g(J)| \geq |g(I)| > A_2 |I|^\gamma > A_3 |J|^\gamma$$

for some A_3.

(3) The proof that Q is embedded proves this claim (3) that Q is a quasi-circle: let $p=h(x)$ and $q=h(y)\in Q$. As before $h(x)$ lies in $T(J_1)$ or its left attached diagonal and $h(y)$ lies in $T(J_2)$ or its right attached diagonal. The arc connecting $h(x)$ and $h(y)$ is contained in the union D of these sets. Then there is a constant K_1 such that $d(x,y)>K_1|D|$.

By (1) the cover $\bigcup g(J)$, $J\in W_k$, satisfies

$$\sum |g(J)|^{1/\gamma}\leq A_1^{1/\gamma}\sum |J|\leq A_1^{1/\gamma}.$$

Hence $\mu_1/\gamma(g(I_0))<\infty$ implying

$$\dim(g(I_0))\leq\frac{1}{\gamma}.$$

Suppose that $\dim(g(I_0))\leq 1/\gamma$. Then $\mu_1/\gamma(g(I_0))=0$. Then there is a cover of $g(I_0)$ by disks B_i such that $\sum |B_i|^{1/\gamma}<\delta$ for any δ. Notice that $\sum |K_iB_i|^{1/\gamma}<K_1^{1/\gamma}\delta$. Consider the cover $\{K_1B_i\}$. Using (3) for each i we pick J_i such that

$$g^{-1}(B_i)\subset J_i\subset \mathrm{cl}(g^{-1}(K_1B_i)).$$

Therefore by (2) we have

$$A_3|J_i|^{\gamma}<|g(J_i)|\leq K_1|B_i|.$$

Hence $|J_i|<(K_1|B_i|/A_3)^{1/\gamma}$. Since $\bigcup J_i$ covers I_0 we have

$$1\leq\sum |J_i|\leq (K_1/A_3)^{1/\gamma}\sum |B_i|^{1/\gamma}.$$

This contradicts the assumption that $\mu_{1/\gamma}=0$. From the preceding paragraph we have $0<\mu_{1/\gamma}(g(I_0))<\infty$. The theorem follows since $\mathrm{cl}(g(I_0))=\Gamma$. ■

Acknowledgment. This research was partially supported by Air Force Grant No. F49620-87-C-0118.

References

[B] P. Blanchard (1984): *Complex analytic dynamics on Riemann sphere.* Bull. Amer. Math. Soc., **11**:85–141.

[HW] G. H. Hardy, E. M. Wright (1974): Introduction to the Theory of Numbers. Oxford: Clarendon Press.

[H1] J. Harrison (1985): *Continued fractals and the Seifert conjecture*, Bull. Amer. Math. Soc., **13**:147–153.

[H2] J. Harrison (1988): *Denjoy fractals.* Topology.

[H3] J. Harrison (1988): *$C^{2+\varepsilon}$ Counterexamples to the Seifert conjecture.* Topology.

[H4] J. Harrison (to appear): *Geometry of algebraic continued fractals.* Proc. London Math. Soc.

[HN] J. HARRISON, A. NORTON (to appear): *Diophantine type and the Denjoy-Koksma inequality.*

[K] H. KESTEN (1966/67): *On a conjecture of Erdös and Szusz related to uniform distribution* mod 1, Arithmetica, **12**:193-212.

[Kh] KHINTCHINE (1963): Continued Fractions, translated by P. Wynn. Groningen: Nordhorff.

[KN] L. KUIPERS, H. NEIDERREITER (1974): Uniform Distribution of Sequences, New York: Wiley.

J. Harrison
Department of Mathematics
University of California
Berkeley
California 94720
U.S.A.

Constr. Approx. (1989) 5: 117–136

CONSTRUCTIVE APPROXIMATION

Porous Surfaces

Claude Tricot

Abstract. In fractal modeling, porous surfaces in the plane are usually described as the residual set E of a packing by connected open domains $\mathscr{C}_n$. In the case where E is nonempty, we investigate the relationships between the dimensionality of E and the geometry of the complementary sets $\mathscr{C}_n$. If they satisfy suitable regularity conditions, then the Bouligand dimension of E is equal to the exponent of convergence of the series $\sum (\text{diam } \mathscr{C}_n)^\alpha$. We give here general conditions to obtain this equality, together with numerous examples and possible ways of developing this theory.

1. Introduction

A mathematical model of porous material can generally be defined as $E = U - \bigcup_{n=1}^{\infty} C_n$ where U, $\mathscr{C}_n$ are open domains such that $\overline{\bigcup_{n=1}^{\infty} \mathscr{C}_n} = \bar{U}$. (Here $\bar{U}$ is the closure of the set U.) For practical applications and references the interested reader may look in [12]. If c_n is the diameter of $\mathscr{C}_n$, the metric properties of the residual set E depend naturally both on the geometrical shape of U and $\mathscr{C}_n$, and on the rate of convergence of the sequence (c_n) to 0. In R, this problem was treated long ago by E. Borel [2] for application to the decimal expansion of a real number, and to the topological properties of the sum of two nowhere dense sets; and a few years later by Besicovitch and Taylor [1] for an approximation of the Hausdorff dimension of E. Indeed, every nowhere dense compact set E on the line is a porous set, with U and $\mathscr{C}_n$ open intervals. If E is the set $\{1/n: n \in N\} \cup \{0\}$, for example, we get $c_n \sim n^{-2}$. If E is the triadic Cantor set, $c_n \sim n^{-\log 3/\log 2}$. Generally, the sequence (c_n) gives rise to a dimensionality notion for E, assuming E has Borel measure 0. Borel defined the *logarithmic rarefaction* of E as

$$(1) \qquad e_B = \inf\left\{\alpha : n^{1-\alpha}\left(\sum_n^{\infty} c_i\right)^{\alpha} \to 0\right\},$$

where the $(\mathscr{C}_n)$ are ordered in such a way that (c_n) is decreasing. The *Besicovitch-Taylor index* is

$$(2) \qquad e_{BT} = \inf\left\{\alpha : \sum_1^{\infty} c_n^{\alpha} < +\infty\right\}.$$

Date received: September 24, 1987. Communicated by Michael F. Barnsley.
AMS classification: 28A75.
Key words and phrases: Fractal dimension, Packing, Porous surfaces.

It happens that these two indices are equivalent, and are also equivalent to a number of other formulations based on the c_n: the simplest is

(3) $$e = \limsup \log n/(-\log c_n),$$

where the c_n are decreasing (see [15] for a general account). The index e, also called exponent of convergence of the series $\sum c_i^\alpha$, by reference to (2), takes the value $\frac{1}{2}$ and $\log 2/\log 3$ respectively on the two previous examples. This is also the value of a well-known definition of dimension, essentially due to G. Bouligand [3], which can be written as follows:

(4) $$\Delta(E) = \inf\{\alpha: r^{\alpha-1}|E(r)| \to 0\},$$

where $|E(r)|$ is the Borel measure of the set of all points at a distance $\leq r$ from E. The quantity $\limsup_{r\to 0} Cr^{\alpha-1}|E(r)|$, with C a normalizing constant, is called the *α-content* of E (Minkowski), and therefore we propose *content dimension* for $\Delta(E)$ [18], although Δ has already been given many other names during its history [15]. The equality

(5) $$e = \Delta(E)$$

is always true if E has measure 0 [10], but of course it fails to be true otherwise, since, in the case of nonzero measure, $\Delta(E)$ is equal to 1.

Thus in $\mathbf{R}$ the choice of the $\mathscr{C}_n$ is canonical and the problem of comparing e and $\Delta(E)$ is solved. We want to investigate the same problem in $\mathbf{R}^2$. For any planar set, (4) becomes

(6) $$\Delta(E) = \inf\{\alpha: r^{\alpha-2}|E(r)| \to 0\},$$

where $|\,|$ is the two-dimensional Borel measure. Given open domains U and $\mathscr{C}_n$, homeomorphic to the open disk, we define e as in (2) or (3). What are the most general conditions on U and $\mathscr{C}_n$ so that equality (5) holds? And, conversely, does there exist, for any compact set E in $\mathbf{R}^2$, a family $(\mathscr{C}_n)$ of complementary sets so that (5) is true? We have already given, in [16, *metric properties of compact sets of measure* 0 *in the plane*], a partial answer to these questions. The present paper is a more complete account, accompanied by examples which illustrate a variety of applications. Our approach is purely geometrical. There exists a statistical approach, where the $\mathscr{C}_n$ are similar copies of the same open set V but with random size and location: we refer the reader to [12] and [23].

Let us state our basic notations and result: *diam* denotes diameter, *dist* denotes the euclidean distance, L is the length of a curve (one-dimensional Hausdorff measure), and $|\,|$ denotes the area of a planar set. For any E, and $r>0$, $E(r) = \{x \in \mathbf{R}^2: \operatorname{dist}(x, E) \leq r\}$. The dimension $\Delta(E)$ is defined by (6), the index $e = e(c_n)$ by (2) or (3). The boundary of E is ∂E.

Theorem 1.1. *Consider an infinite collection $U = \mathscr{C}_0, \mathscr{C}_1, \mathscr{C}_2, \ldots$ of sets homeomorphic to the open disk S_2, such that $\mathscr{C}_n \subset U$ if $n \geq 1$. We assume that the curve $\partial\mathscr{C}_n$ is rectifiable for $n \geq 0$, of length $L(\partial\mathscr{C}_n) = l_n$. Note that $\bar{U} - \bigcup_1^\infty \mathscr{C}_n$ is not empty. A residual set E is any nonvoid Borel set such that*

$$U - \bigcup_1^\infty \bar{\mathscr{C}}_n \subset E \subset \bar{U} - \bigcup_1^\infty \mathscr{C}_n.$$

Let $c_n = \operatorname{diam} \mathscr{C}_n$. *We are interested in the six following properties*:

(i) $c_n^2 = \mathcal{O}(|\mathscr{C}_n|)$.
(ii) $\operatorname{dist}(\mathscr{C}_n, E) = \mathcal{O}(c_n)$.
(iii) *For any* $x \in U$, *the number of sets* $\mathscr{C}_n$ *containing* x *is smaller than a fixed number* N.
(iv) $|\bigcup \mathscr{C}_n| = |U|$.
(v) $l_n = \mathcal{O}(c_n)$.
(vi) *If* U_n *is the union of all* $\mathscr{C}_i$ *such that* $c_i \geq c_n$, *then the number of connected components of* $U - \bar{U}_n$ *is* $\mathcal{O}(n)$.

A. *If* (i), (ii), *and* (iii) *hold, then* $e \leq \Delta(E)$.
B1. *If* (iv) *and* (v) *hold, then* $\Delta(E) \leq \max(e, 1)$.
B2. *If* (iv), (v), *and* (vi) *hold, then* $\Delta(E) \leq e$.

The proof is given in Sections 3, 4, and 6, together with examples showing that if all but one of conditions (i)–(vi) are verified the equality $e = \Delta(E)$ may not be true. In the proof we assume that the $\mathscr{C}_n$ are ordered as c_n is decreasing. We first give, in Section 2, some technical lemmas on the exponent of convergence of the series $\sum c_n^\alpha$. Sections 5 and 7 are devoted to an eclectic choice of applications, and Section 8 to a study of a particular class of sets: the self-similar sets. For them we show that it is possible to find a complementary family $(\mathscr{C}_n)$ which is also, in some sense, self-similar. Section 9 shows different ways to weaken the hypothesis of Theorem 1.1, without giving any detailed proof to save space, the main arguments being essentially the same. An attempt to write the most general version of Theorem 1.1 would be very cumbersome: our statement was inspired by the subsequent applications, and also by our search for simplicity, but variants do exist. In Section 10 we discuss the problem in $\mathbf{R}^p$, $p \geq 3$, which leads to awkward difficulties. We show that any compact set E may be considered as the complement of a Whitney family of cubes which satisfy the hypothesis of Theorem 1.1. Finally, we could ask for the interest and interpretation of e in the case where E has nonzero measure (condition (iv) is not verified). This problem has already been investigated: see [9] and [18] for applications, and [19] for a general account.

2. Exponents of Convergence

The following two propositions constitute the fundamental technique for the proof of Theorem 1.1, and they are of great use for most problems of packing. Proposition 2.1 is a generalization of results in [22] or [15].

Proposition 2.1. *Let* (c_n) *be a decreasing sequence, tending to* 0, *and* a, b, σ *three positive real numbers,* $\sigma \neq 0$. *We denote*

$$e = e(c_n) = \limsup \log n / (-\log c_n).$$

Assume $\sum_1^\infty n^b c_n^\sigma < +\infty$, *and let*

$$e^* = e^*(c_n; a, b, \sigma) = \inf\left\{ \alpha : n^a c_n^{\alpha - \sigma} \sum_n^\infty i^b c_i^\sigma \to 0 \right\}.$$

Then $e^* = (a + b + 1)e$.

Proof. If $c'_n = c_n^\sigma$, then $e = \sigma \limsup \log n/(-\log c'_n)$, and $e^*(c_n) = \inf\{\alpha: n^a c_n'^{(\alpha/\sigma)-1} \sum_n^\infty i^b c'_n \to 0\} = \sigma e^*(c'_n)$. We may therefore assume that $\sigma = 1$. We observe that $\sum n^b c_n < +\infty$ implies $c_n = \mathcal{O}(n^{-b-1})$ (use the fact that $\sum_n^{2n} i^b c_i \to 0$), so that $e \le 1/(b+1)$.

1. Let $\alpha > (a+b+1)e$. We have $c_n = o(n^{-(a+b+1)/\alpha})$. Two cases:

(a) If $e \ge 1/(a+b+1)$, then $\alpha > 1$. If $e = 1/(b+1)$, then $n^a c_n^{\alpha-1} = o(n^\gamma)$, where $\gamma = a - (\alpha-1)(a+b+1)/\alpha < 0$ since $\alpha > (a+b+1)/(b+1)$. Therefore $e^* \le \alpha$. If $e < 1/(b+1)$, we consider $\sum_n^\infty i^b c_i = o(\sum_n^\infty i^\beta)$, where $\beta = b - (a+b+1)/\alpha < -1 + (1/e) - (a+b+1)/\alpha$. If α is close enough to $(a+b+1)e$, the series $\sum_n^\infty i^b c_i$ is $o(n^{1+\beta})$. Using $\alpha - 1 > 0$ we get $n^a c_n^{\alpha-1} \sum_n^\infty i^b c_i = o(n^\gamma)$, where $\gamma = a - ((\alpha-1)(a+b+1)/\alpha) + 1 + \beta = 0$. Therefore $e^* \le \alpha$.

(b) If $e < 1/(a+b+1)$, take α and ε such that $(a+b+1)e < \alpha < \alpha + \varepsilon < 1$. Then

$$n^a c_n^{\alpha+\varepsilon-1} \sum_n^\infty i^b c_i \le \sum_n^\infty i^{a+b} c_i^{\alpha+\varepsilon} = o\left(\sum_n^\infty i^\beta\right),$$

where $\beta = a + b - (\alpha+\varepsilon)(a+b+1)/\alpha < -1$. Since $\sum_n^\infty i^\beta$ converges, we get $e^* \le \alpha + \varepsilon$. We have proved that $e^* \le (a+b+1)e$.

2. Let $\alpha > e^*$. We have $\sum_n^\infty i^b c_i = o(n^{-a} c_n^{1-\alpha})$. Applying Dini's theorem, for all $\beta < 1$ the series $\sum_1^\infty n^b c_n (\sum_n^\infty i^b c_i)^{-\beta}$ is convergent, so that

$$\sum_1^\infty n^{b+\beta a} c_n^{1+\beta(\alpha-1)} < +\infty.$$

We deduce $c_n^{1+\beta(\alpha-1)} = \mathcal{O}(n^{-1-b-\beta a})$ and $e \le (1+\beta(\alpha-1))/(\beta a + b + 1)$. Let $\beta \to 1$ and $\alpha \to e^*$ to obtain the desired result. This completes the proof. ∎

We now need to study the divergent case:

Proposition 2.2. *Let (c_n) be a decreasing sequence, tending to 0, and a, b, σ three positive real numbers, $\sigma \ne 0$. We denote*

$$e = e(c_n) = \limsup \log n/(-\log c_n).$$

Assume $\sum_1^\infty n^b c_n^\sigma = +\infty$, and let

$$e^{**} = e^{**}(c_n;\, a, b, \sigma) = \inf\left\{\alpha: n^a c_n^{\alpha-\sigma} \sum_1^n i^b c_i^\sigma \to 0\right\}.$$

Then $e^{**} = (a+b+1)e$.

Proof. As in Proposition 2.1 we assume $\sigma = 1$. The divergence of the series $\sum n^b c_n$ implies $e \ge 1/(b+1)$.

1. Let $\alpha > (a+b+1)e$, so that $\alpha > 1$. Since $n^b c_n = o(n^{b-(a+b+1)/\alpha})$, we have $\sum_1^n i^b c_i = o(n^\beta)$ where $\beta = 1 + b - (a+b+1)/\alpha$. Using $\alpha - 1 > 0$, $n^a c_n^{\alpha-1} \sum_1^n i^b c_i = o(n^\gamma)$, where $\gamma = a - (\alpha-1)(a+b+1)/\alpha + \beta = 0$. Hence $e^{**} \le \alpha$.

2. Let $\alpha > e^{**}$. Then $\sum_1^n i^b c_i = o(n^{-a} c_n^{1-\alpha})$. Since $n^{b+1} c_n \sim c_n \sum_1^n i^b \leq \sum_1^n i^b c_i$, we get $c_n = o(n^{-a-b-1} c_n^{1-\alpha})$, and $c_n^\alpha = o(n^{-a-b-1})$. Therefore $e \leq \alpha/(a+b+1)$. ■

Remark 2.3. The proofs of Propositions 2.1, part 1(a), and 2.2, are still valid if $a<0$: all we need is $a+b\geq 0$. These propositions could be written in a more general context but we do not need it here.

3. Part A of Theorem 1.1

We begin with examples of families $(\mathscr{C}_n)$ satisfying all conditions of the theorem except for (i) and (ii), and such that $\Delta(E)<e$. We take $E=\bar{U}-\bigcup_{i=1}^{\infty}\bar{\mathscr{C}}_n$. Let $J=]0,1[$, $I=[0,1]$.

Example 3.1. Take $U=J\times J$ and $\mathscr{C}_n=]1/(n+2), 1/n[\times J$. The set E is the segment $I\times\{0\}$, and conditions (ii)-(vi) are verified (take $N=2$ for (iii)), but not (i). We have $\Delta(E)=1$, and, since $c_n>1$, $e=+\infty$. Of course it is possible to arrange this example so that c_n tends to 0, and $e=2$.

Example 3.2. Take $U=J\times J$, and for $\mathscr{C}_n$ the family of all rectangles of the type $]1/(n+2), 1/n[\times]k/n(n+1), (k+2)/n(n+1)[$ for $k=0,\ldots,n(n+1)-2$ and $n\geq 1$. Again $E=I\times\{0\}$ and $N=2$. Since there are $n(n+1)-2$ rectangles of diameter $\sim n^{-2}$, the rank at which those diameters occur in the sequence (c_n) has the same order of growth as n^3. Therefore $e=\frac{3}{2}$, and $\Delta(E)=1$. Here property (ii) is not verified: $\operatorname{dist}(\mathscr{C}_n, E)\sim c_n^{1/2}$.

To show the necessity of a condition such as (iii) it suffices to notice that by using as many copies as we want of the same $\mathscr{C}_n$, for infinitely many n, we can artificially increase the value of e.

Proof of Part A. Suppose the c_n are decreasing. Condition (ii) implies that, for some constant A, and any $r<1$, $E(Ar)$ contains all $\mathscr{C}_i$ of rank $\geq n+1$, where $c_{n+1}\leq r<c_n$. Therefore

$$\left|\bigcup_{n+1}^{\infty}\mathscr{C}_i\right|\leq|E(Ar)|. \tag{7}$$

Condition (iii) implies

$$\sum_{n+1}^{\infty}|\mathscr{C}_i|\leq N\left|\bigcup_{n+1}^{\infty}\mathscr{C}_i\right|. \tag{8}$$

Since, for some constant B, (i) implies

$$c_i^2\leq B|\mathscr{C}_i| \qquad \text{for all } i, \tag{9}$$

(7), (8), and (9) give

$$\sum_{n+1}^{\infty} c_i^2\leq BN|E(Ar)|. \tag{10}$$

This shows in particular that $e \leq 2$. We know that $\Delta(E) \leq 2$ (as any planar set). If $\Delta(E) = 2$ there is nothing to prove. Otherwise, take $\Delta(E) < \alpha < 2$, and deduce from (10) that

$$c_n^{\alpha-2} \sum_{n+1}^{\infty} c_i^2 \leq BNr^{\alpha-2}|E(Ar)|.$$

With $\varepsilon = Ar$, the right member is equivalent to $\varepsilon^{\alpha-2}|E(\varepsilon)|$ and tends to 0. The left member can be written as $c_n^{\alpha-2} \sum_n^{\infty} c_i^2 - c_n^{\alpha}$ and tends to 0 as well. Now apply Proposition 2.1 with $a = b = 0$ and $\sigma = 2$ to get $e \leq \alpha$.

4. Part B1 of Theorem 1.1

Example 4.1. Let us show that, even if $\cup \mathscr{C}_n$ is dense in U, such a condition as (iv) is necessary: take a nowhere dense compact set F in I, of measure $\frac{1}{2}$, whose complement is a disjoint family of 3^n triadic intervals of length 9^{-n}. The image of a triadic interval of length 9^{-n} by a suitable Peano curve $f\colon I \to I \times I$ is a triadic square of side 3^{-n}. Take $U = J \times J$, and $E = f(F)$: since f preserves the measure, $|E| = \frac{1}{2}$, so that $\Delta(E) = 2$. Taking for $(\mathscr{C}_n)$ the images of the complementary intervals of F, we get, for all n, 3^n squares of side 3^{-n}, so that $e = 1$.

Example 4.2. Let us now exhibit a family $(\mathscr{C}_n)$ which does not satisfy (v). Let Γ be the Von Koch curve ("snowflake curve") constructed on a segment AB of length 1 (for a definition and pictures of this well-known curve, see, for example, [12]). We have $\Delta(\Gamma) = \log 4/\log 3$. The convex envelope of Γ is a triangle ABC, with two equal sides $AC = BC$. We take for U the inside of this triangle. Denote by D_1, D_2, D_3 the insides of the domains bounded by $AB \cup \Gamma$, $BC \cup \Gamma$, and $CA \cup \Gamma$, respectively. Since D_2 and D_3 are similar to D_1, with similarity ratio $3^{-1/2}$, we will only consider the packing of D_1: similar constructions may be performed in D_2, D_3 and that does not change the value of e. The curve Γ is the limit of a sequence (Γ_k) of polygonal curves in $\bar{D}_1$, each Γ_k being composed of 4^k segments of length 3^{-k}. The inside of the domain bounded by $\Gamma_k \cup \Gamma_{k+1}$ is constituted of 4^k equilateral triangles of side 3^{-k-1}. If we take for family $(\mathscr{C}_n)$ the collection of all such triangles, for all $k \geq 1$, we get the expected value $e = \log 4/\log 3$. Being given a strictly increasing sequence (n_k) of integers, we will rather consider the connected domains bounded by $\Gamma_{n_k} \cup \Gamma_{n_{k+1}}$ (finite unions of triangles), in order to get more complicated shapes: the perimeter of such figures will still be finite, but if n_k grows very quickly, property (v) will not be satisfied and the value of e will be smaller. Let us put $n_k = i$, $n_{k+1} = j$ for simplicity. The linear set $\Gamma_i \cup \Gamma_j$ encloses $a(i, j) = 2^{j+i} - 2^{2i}$ open, connected domains of diameters varying between 3^{-i} and 3^{-j}, the largest having perimeter $3^{-i-1} + 2^{-2i-1}(4/3)^j$. The shaded area of Fig. 1 represents the case $i = 1, j = 4$. There are 28 connected components, the largest having perimeter 41/81.

Suppose (n_k) is equivalent to $\exp(\exp k)$. Denote by $\mathscr{X}_k$ the family of the connected domains bounded by $\Gamma_{n_k} \cup \Gamma_{n_{k+1}}$. Since the diameter of any figure in $\mathscr{X}_k$ is larger than that of any figure in $\mathscr{X}_{k+1}$, and since $\log \sum_1^k a(n_i, n_{i+1}) \sim n_{k+1} \log 2$,

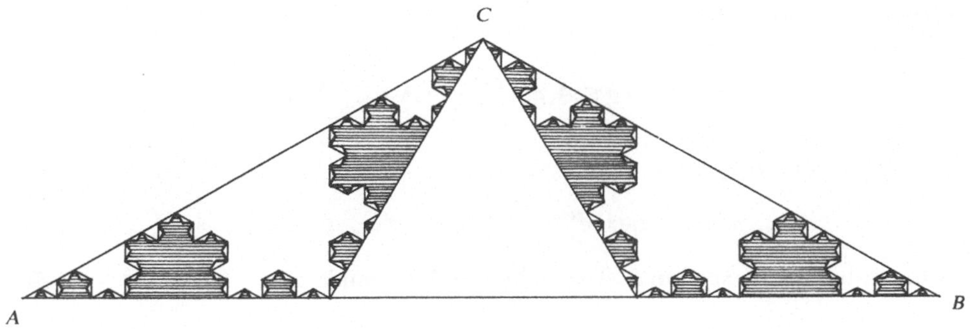

Fig. 1

we get $e = \log 2/\log 3$. Indeed, (v) is not verified, since the largest perimeter in $\mathscr{X}_k$ tends to $+\infty$ with k.

Lemma 4.3. *If C is a simple, rectifiable arc in the plane, and $\varepsilon < L(C)$, then*

$$|C(\varepsilon)| \le 8\pi\varepsilon L(C).$$

Proof. Let $M_\varepsilon(C)$ be the maximum number of points in C at distance $\ge \varepsilon$ from each other. If $M_\varepsilon(C) = 1$, we get $\varepsilon M_\varepsilon(C) \le L(C)$. If $M_\varepsilon(C) \ge 2$, $\varepsilon(M_\varepsilon(C) - 1) \le L(C)$, so that $\varepsilon M_\varepsilon(C) \le 2L(C)$. Since $C(\varepsilon)$ is included in $M_\varepsilon(C)$ balls of radius 2ε, it follows that

$$|C(\varepsilon)| \le 4\pi\varepsilon^2 M_\varepsilon(C) \le 8\pi\varepsilon L(C).$$

Of course this constant 8π is not the smallest possible. ■

Proof of B1. Suppose c_n is decreasing. Let us take $E = \bar{U} - \bigcup_1^\infty \mathscr{C}_n$. We may assume that $e < 2$. This implies $\sum c_n^2 < +\infty$. For all $n \ge 0$, $r > 0$, $E(r) \cap \mathscr{C}_n \subset \partial\mathscr{C}_n(r)$. We fix $r > 0$, and take n such that $c_{n+1} < r \le c_n$. Lemma 4.3 implies

$$\forall i, 0 \le i \le n: \qquad |E(r) \cap \mathscr{C}_i| \le 8\pi r l_i.$$

Condition (v) gives that, for such i, $|E(r) \cap \mathscr{C}_i| \le A_1 r c_i$ for some constant A_1. Finally, from condition (iv) it follows that

$$\begin{aligned} |E(r)| &\le \sum_0^n |E(r) \cap \mathscr{C}_i| + \sum_{n+1}^\infty |\mathscr{C}_i| \\ &\le A_1 r \sum_0^n c_i + A_2 \sum_{n+1}^\infty c_i^2. \end{aligned}$$

For all α, $\max(e, 1) < \alpha < 2$,

$$r^{\alpha-2}|E(r)| \le A_1 r^{\alpha-1} \sum_0^n c_i + A_2 r^{\alpha-2} \sum_{n+1}^\infty c_i^2.$$

Since $\alpha-1>0, \alpha-2<0$, we get

$$r^{\alpha-2}|E(r)| \leq A_1 c_n^{\alpha-1} \sum_{0}^{n} c_i + A_2 c_{n+1}^{\alpha-2} \sum_{n+1}^{\infty} c_i^2.$$

This proves the inequality

$$\Delta(E) \leq \max(e^{**}(c_n; 0, 0, 1), e^*(c_n; 0, 0, 2))$$

in the notations of Section 2. The right member equals e. ■

5. Applications of A and B1

Application 5.1. Consider the apollonian packing, constructed in a curvilinear triangle U by packing disjoint disks of maximal size (Fig. 2). If (c_n) is the sequence of the radii, the exponent of convergence has been called the "packing constant" [13] (see also [20] and [21]). This is a simple example of complementary sets verifying (i)-(v). Let $E = \bar{U} - \bigcup_1^\infty \mathscr{C}_n$. Since E contains arcs, $\Delta(E) \geq 1$, so that, from A and B1, $\Delta(E) = e$. An approximate value of e is 1.306 [5]. A more difficult problem was to show that e is also the Hausdorff dimension of E [6], [17].

Application 5.2. We want to generalize 5.1 by considering other geometrical shapes. What happens if we pack a given domain U by ellipses, triangles, or even nonconvex bodies? To make sure that condition (ii) is verified, and also that $\Delta(E) \geq 1$, we will give a condition on the boundaries.

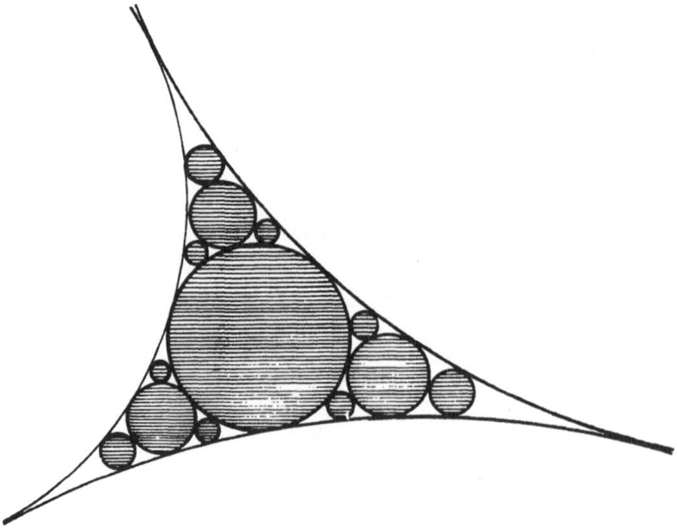

Fig. 2

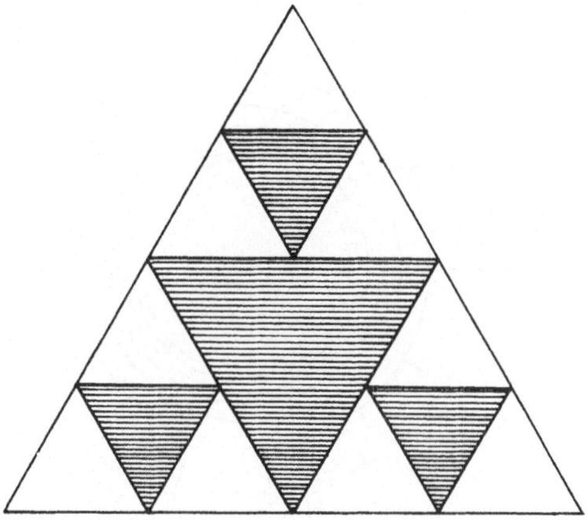

Fig. 3

Proposition 5.3. *Let U, V be open domains, homeomorphic to S_2, with rectifiable boundaries. Let $\mathscr{C}_n$ be an infinite sequence of disjoint, similar copies of V, included in U, such that, for all $n \neq k$, $L(\partial\mathscr{C}_n \cap \partial\mathscr{C}_k) = 0$. Let d_n be the diameter of the largest disk in $U - \bigcup_{i=1}^{n-1} \mathscr{C}_i$. We construct the family $(\mathscr{C}_n)$ in such a way that $d_n = \mathcal{O}(c_n)$. Then $e = \Delta(E)$.*

These hypothesis are easily verified if V is strictly convex, and if, for all n, $\mathscr{C}_n$ is the largest possible copy of V in $U - \bigcup_{i=1}^{n-1} \mathscr{C}_i$. But $\mathscr{C}_n$ may not be the largest, for example, see Fig. 3, representing the well-known packing of an equilateral triangle by inverted triangles: here $c_n = d_n$, but the largest possible C_1 in U would be U itself. The residual set is such that $e = \Delta(E) = \log 3/\log 2$.

Proof of Proposition 5.3. Conditions (i), (iii), and (v) of Theorem 1.1 are clearly verified. Since $L(E \cap \partial\mathscr{C}_n) = l_n$ for all $n > 0$, (ii) is verified and $\Delta(E) \geq 1$. All we have to check is that $|U| = |\bigcup \mathscr{C}_n|$. Suppose $d_n \leq Kc_n$ for all n. The set $\bigcup \mathscr{C}_n$ is everywhere dense in U. It is enough to prove that, B_n being a ball centered in $\mathscr{C}_n$, of radius $(K+1)c_n$, we have for all n

$$U - \bigcup_{i=1}^{n} \bar{\mathscr{C}}_i \subset \bigcup_{n+1}^{\infty} B_i.$$

If we take any ball B included in $U - \bigcup_{i=1}^{n} \bar{\mathscr{C}}_i$, and if N is the smallest integer such that $B \cap \mathscr{C}_N \neq \emptyset$, then $N > n$. The fact that $B \cap \mathscr{C}_i = \emptyset$ for all $1 \leq i \leq N-1$ implies that diam $B \leq d_N$, so diam $B \leq Kc_N$: therefore $B \subset B_N$.

Remark 5.4. If we replace, in Proposition 5.3, the set V by a family (V_n), verifying (i) and (v), we may take for $\mathscr{C}_n$ a copy of V_n and the result is still true. Also the condition on the boundaries may be weakened, by using part B2 of Theorem 1.1: see Section 6.

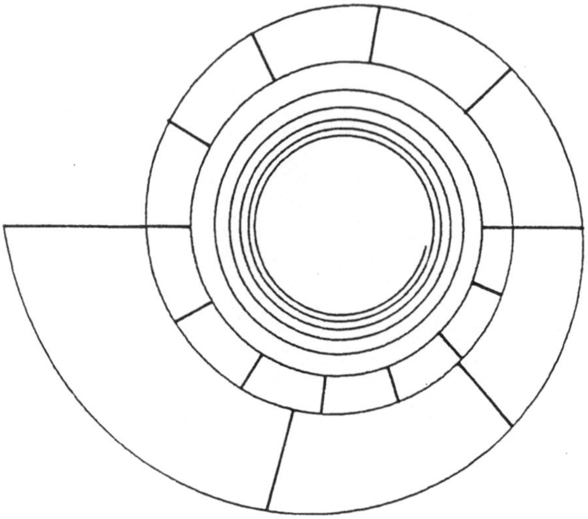

Fig. 4

Application 5.5. Consider the spiral Γ defined in polar coordinates by $\rho = \theta^{-\beta}$, $0<\beta\leq 1$, for all $\theta \geq \pi$. Let $\theta_n = \pi\sqrt{n}$. We take an open domain U consisting of all points (ρ, θ) such that $\theta > \pi$ and $\rho < \theta^{-\beta}$, and open domains $\mathscr{C}_n$ such that $\theta_n < \theta < \theta_{n+2}$ and $(\theta+2\pi)^{-\beta} < \rho < \theta^{-\beta}$ (Fig. 4 represents the case $\beta = \frac{1}{2}$). Take $E = \bar{U} - \bigcup_{i=1}^{\infty} \mathscr{C}_i = \Gamma$. Conditions (ii)-(v) are easily verified. Condition (i) follows from the convergence of the ratio $(\theta_{n+2} - \theta_n)\theta_n^{-\beta}/(\theta_n^{-\beta} - (\theta_n + 2\pi)^{-\beta})$ to a finite limit: this ensures that the four curvilinear sides of $\mathscr{C}_n$ have equivalent lengths. Since $\Delta(E) \geq 1$, Theorem 1.1 implies $\Delta(E) = e$. The value of e is given by $c_n \sim \theta_n^{-\beta} - (\theta_n + 2\pi)^{-\beta} \sim n^{-(\beta+1)/2}$: $e = 2/(\beta+1)$. A study of spirals in a more general context needs a careful analysis [8], using the exponents $e^*(c_n; 0, 0, 1)$, $e^*(c_n; 1, 0, 1)$, $e^{**}(c_n; 0, 1, 1)$, and $e^{**}(c_n; 1, 0, 0)$.

6. Part B2 of Theorem 1.1

First let us see why we need such a condition as (vi):

Example 6.1. Take $U = J \times J$, (a_n) a sequence of positive numbers such that $a_0 = 1$ and $a_{n+1} < a_n/2$, and E_n the union of 2^n disjoint, open segments in $I \times \{0\}$ of length a_n, such that each of them contains two segments of E_{n+1}. Let $n(0) = 0$, and $(n(k))$ be a strictly increasing sequence of integers. We denote by $\mathscr{C}_1$ the open set $U - (\bar{E}_{n(1)} \times [0, a_{n(1)}])$. Generally we form the family $(\mathscr{C}_n)$ with all connected domains in $(E_{n(k)} \times]0, a_{n(k)}[) - (\bar{E}_{n(k+1)} \times [0, a_{n(k+1)}])$ for all $k \geq 0$. For every k there are $2^{n(k)}$ such domains, of diameter $a_{n(k)}\sqrt{2}$, of inside diameter $a_{n(k)} - a_{n(k+1)} \geq a_{n(k)}/2$, and of perimeter equivalent to $4a_{n(k)} + 2^{n(k+1)-n(k)} a_{n(k+1)} \leq 6a_{n(k)}$. The residual set $E = \overline{U - \bigcup_{n=1}^{\infty} \mathscr{C}_n}$ is equal to $\bigcap_{n=1}^{\infty} \bar{E}_n$: it is a nowhere dense, perfect set in $I \times \{0\}$. Since the domains above are at a distance $\leq a_{n(k+1)}$ from E, condition (ii) is verified. Condition (iii) is also verified, since the $\mathscr{C}_n$ are disjoint. Finally (i), (iv), and (v) are easy to check. Now let $m(k)=\sum_{i=1}^{k} 2^{n(i)}$: the number of connected components of $U-\bigcup_{i=1}^{m(k)} \bar{\mathscr{C}}_i$ is equivalent to $2^{n(k+1)}$. We may choose (a_n) and $n(k)$ so that

(a) $\limsup \log 2^n/(-\log a_n)=1$,

(b) $\log 2^{n(k)}/(-\log a_{n(k)})\to 0$, and

(c) $\liminf m(k)2^{-n(k+1)}\to 0$.

Property (a) implies $\Delta(E)=1$, (b) $e=0$, and (c) condition (vi) is not verified. Actually, property (c) is implied by (a) and (b) (if $n_{(k+1)}/n(k)$ is bounded, (a) and (b) are incompatible; if $n(k+1)/n(k)$ is not bounded, then (c) is true). Property (b) implies that the symmetric dimension $\delta(E)=\sup\{\alpha: r^{\alpha-2}|E(r)|\to+\infty\}$ has value 0. We do not have any example of a family $(\mathscr{C}_n)$ verifying (i)-(v), but failing to verify (vi), and such that $\delta(E)=\Delta(E)$.

For the proof of B2 we need the following result, which is of interest in itself:

Lemma 6.2. *Let $U=\mathscr{C}_0, \mathscr{C}_1,\dots$ be a family of open domains homeomorphic to S_2, with rectifiable boundaries of length l_n, such that $\mathscr{C}_n\subset U$ and $|U|=|\bigcup_1^\infty \mathscr{C}_n|$. Then*

$$l_0\leq\sum_1^\infty l_n.$$

Proof. We assume that $\mathscr{C}_n$ is an infinite family, and $\sum_1^\infty l_n<+\infty$. Denote by F_n the boundary $\partial\mathscr{C}_n$ of $\mathscr{C}_n$. Let $E=\overline{U-\bigcup_1^\infty \mathscr{C}_n}$ be the residual set: $|E|=0$. In $\bar{U}$, E is nowhere dense and $\bigcup_1^\infty \mathscr{C}_n$ is dense. We first define a family G_n of arcwise connected, compact sets in $\bar{U}$: take $G_1=F_1$, $G_{1,1}=F_1$, $G_{1,2}=\varnothing$. By induction: assume G_n is constructed as a subset of a union of rectifiable arcs, and can be decomposed as $G_n=G_{n,1}\cup G_{n,2}$, where $\bigcup_{i=1}^n F_i\subset G_{n,1}\subset\bigcup_{i=1}^\infty F_i$, and $L(G_{n,2})=0$. We construct G_{n+1} as follows:

(a) If $F_{n+1}\cap G_n\neq\varnothing$, take $G_{n+1}=G_n\cup F_n$. It is clear that G_{n+1} will have the same properties as G_n.

(b) If $F_{n+1}\cap G_n=\varnothing$, since $|E|=0$ we can find a rectifiable simple arc C in U with one extremity in G_n, the other in F_{n+1}, such that $L(C\cap E)=0$. Such an arc necessarily meets some $\bar{\mathscr{C}}_k$ for $k\geq n+2$. Let us call (n_k) the sequence of indices k for which $\bar{\mathscr{C}}_k\cap C\neq\varnothing$. Assume moreover that, if $i\neq j$, $\bar{\mathscr{C}}_{n_i}\cap C$ is not included in $\bar{\mathscr{C}}_{n_j}\cap C$ (take a restricted sequence if necessary). Now, writing $F'_k=F_{n_k}$, we define

$$C'=(C-\bigcup_0^\infty\mathscr{C}_n)\cup(\cup\ F'_k)$$

and

$$G_{n+1}=G_n\cup C'.$$

Since C' is included in a union of rectifiable arcs, so is G_{n+1}. Moreover, $F_{n+1}\subset C'$, so we may write $G_{n+1}=G_{n+1,1}\cup G_{n+1,2}$ where $G_{n+1,1}=G_{n,1}\cup(\cup\ F'_k)$ contains $\bigcup_{i=1}^{n+1} F_i$, is contained in $\bigcup_1^\infty F_i$, and $G_{n+1,2}=G_{n,2}\cup(C\cap E)$ has linear measure

0. In some sense C' is a way to go from G_n to F_{n+1} passing through the boundaries of the $\mathscr{C}_n$. To complete the induction, all we have to check is (b1) C' is arcwise connected, and (b2) C' is compact.

(b1) We obtained C' by replacing a finite or countable union of subarcs of C by arcs of the same extremities (the F'_k). This shows that, for every pair (P, Q) of points in C', there exists in C' a continuous path joining P to Q.

(b2) Let us show that $\overline{\bigcup F'_k} = C'$. Take a sequence (x_j) of points in $\bigcup F'_k$, converging to a limit x. Let k_j be such that $x_j \in F'_{k_j}$. If $\{k_j\}$ is a finite set, then an infinity of x_j belongs to the same F'_k for some k, and $x \in F'_k$. Otherwise, let us assume that (k_j) is strictly increasing: since $k_j \to +\infty$, $\text{dist}(x_j, C) \to 0$, and $x \in C$. It is impossible that $x \in \mathscr{C}_n$ from some n, indeed, $\mathscr{C}_n$ is open, and would include an infinity of F'_k; this is impossible by construction of the F'_k. Therefore, $x \in C - \bigcup_0^\infty \mathscr{C}_n \subset C'$, and C' is compact.

The G_n form a sequence of continua, included in a bounded domain, and with bounded lengths: indeed, $L(G_n) = L(G_{n,1}) \le \sum_1^\infty l_i$. We deduce (Gotab 1929) that a subsequence converges to a continuum G such that $L(G) \le \sum_1^\infty l_i$. By construction $\bigcup_1^\infty f_n \subset G$, and, since G is compact, $\overline{\bigcup_1^\infty F_n} \subset G$. But $\bigcup \mathscr{C}_n$ is dense in $\bar{U}$, so that $F_0 \subset \overline{\bigcup F_n}$. Therefore $l_0 \le L(G)$, and the proof is completed. ∎

Proof of B2. If $e \ge 1$, part B1 suffices. Let us assume that $e < 1$, which implies $\sum c_n < +\infty$. Assume c_n is decreasing. Let $r > 0$ and n be such that $c_{n+1} < r \le c_n$. We denote by $U_1, \ldots, U_M$ the connected components of $U - \bigcup_1^n \bar{\mathscr{C}}_i$. Their boundaries are of finite linear measure. Assume that, for some $N \le M$, we have $i \le N \Leftrightarrow L(\partial U_i) > r$. Lemma 6.2 implies that

$$L(\partial U_i) \le \sum l_k,$$

where the sum is extended to all $\mathscr{C}_k$, $k \ge n+1$, included in U_i. Consider the inclusion

$$E(r) \subset \left(\bigcup_{i=1}^{N} (\partial U_i)(r)\right) \cup \left(\bigcup_{r=N+1}^{M} (\partial U_i)(r)\right) \cup \left(\bigcup_{i=n+1}^{\infty} \mathscr{C}_i\right) \cup E \cup (\partial U(r)).$$

Lemmas 4.3 and 6.2 give

$$\sum_{i=1}^{N} |\partial U_i(r)| = \mathscr{O}\left(r \sum_{k=n+1}^{\infty} l_k\right),$$

the right member being equivalent to $r \sum_{k=n+1}^{\infty} c_k$ by condition (v). We also have

$$\sum_{i=N+1}^{M} |(\partial U_i)(r)| \le 3\pi M r^2,$$

the right member being $\mathscr{O}(nr^2)$ from condition (vi). Finally, $|E| = 0$ by (iv), $\sum_{i=n+1}^{\infty} |\mathscr{C}_i| = \mathcal{O}(\sum_{i=n+1}^{\infty} c_i^2)$, and $|\partial U(r)| = \mathcal{O}(r)$. We get, for every $0<\alpha<1$,

$$r^{\alpha-2}|E(r)| = \mathcal{O}\left(r^{\alpha-1}\sum_{n+1}^{\infty} c_k + nr^{\alpha} + r^{\alpha-2}\sum_{n+1}^{\infty} c_i^2\right)$$

$$= \mathcal{O}\left(c_{n+1}^{\alpha-1}\sum_{n+1}^{\infty} c_k + nc_n^{\alpha} + c_{n+1}^{\alpha-2}\sum_{n+1}^{\infty} c_i^2\right).$$

It follows that

$\Delta(E) \le \max(e^*(c_n; 0, 0, 1), e^{**}(c_n; 0, 0, 0), e^*(c_n; 0, 0, 2))$. From the results in Section 2, all three numbers equal e. ■

7. Applications of A and B2

Application 7.1. Condition (vi) is implied by conditions (i) and (iii) if the $\mathscr{C}_n$, for all $n \ge 1$, are convex bodies, and U is the complement in the plane of a finite union of convex domains (U may be a polygon for example). Indeed, let us assume that the c_n are decreasing. Conditions (i) and (iii) imply that there exists an integer K such that, for all n, $\mathscr{C}_n$ cannot meet more than K other $\mathscr{C}_i$, for $0 \le i \le n-1$. Let us call $V_1, \ldots, V_M$ the connected components of $U_n = U - \bigcup_{i=1}^{n} \mathscr{C}_i$ whose boundary meets $\partial\mathscr{C}_n$. Each V_i is included in a connected component $\mathscr{V}_i$ of U_{n-1} (the $\mathscr{V}_i$ are not necessarily distinct). Let us put $\varepsilon_i = 0$ if V_i is the only new connected component in $\mathscr{V}_i$, and $\varepsilon_i = 1$ otherwise, i.e., if $\mathscr{C}_n$ creates at least two connected components in $\mathscr{V}_i$. To verify (vi), all we prove is that $\sum_{i=1}^{M} \varepsilon_i$ is bounded by some constant c. We assume $M > 1$. Using the convexity, we proceed as follows:

Let C_j be the largest arc in $\partial\mathscr{C}_n$ whose extremities x_j and y_j belong to ∂V_j. The arcs $C_1, \ldots, C_M$ are disjoint. The points x_j, y_j are also in $\bigcup_{i=0}^{n-1} \partial\mathscr{C}_i$. If there exists $i \le n-1$ such that x_j and $y_j \in \partial\mathscr{C}_i$, then $\varepsilon_j = 0$. Otherwise, we can find a pair (a_j, b_j), $0 \le a_j < b_j \le n-1$, such that $x_j \in \partial\mathscr{C}_{a_j}$ and $y_j \in \partial\mathscr{C}_{b_j}$, or $x_j \in \partial\mathscr{C}_{b_j}$ and $y_j \in \partial\mathscr{C}_{a_j}$. If the intersection $\mathscr{C}_{a_j} \cap \mathscr{C}_{b_j} \cap \mathscr{C}_n \ne \varnothing$, then again $\varepsilon_j = 0$. We deduce that, given three integers j, k, l, $1 \le j < k < l \le M$, the equalities $(a_j, b_j) = (a_k, b_k) = (a_l, b_l)$ imply $\varepsilon_j = \varepsilon_k = \varepsilon_l = 0$. In conclusion, every pair (a_j, b_j) as above can be associated to 0, 1, or at most 2 arcs C_k such that $\varepsilon_k = 1$. Since there are no more than $K(K-1)$ such pairs, it follows that

$$\sum_{1}^{M} \varepsilon_j \le c = 2K(K-1).$$

In this case proposition 5.3 may be simplified as follows:

Proposition 7.2. *Let V be a bounded, open, convex set and U a bounded domain, the complement of a finite, connected union of convex domains. We take for $\mathscr{C}_n$ an infinite sequence of disjoint, similar copies of V in U, and assume $d_n = \mathcal{O}(c_n)$. Then $e = \Delta(E)$.*

Application 7.3. This proposition applies to the packing of an equilateral triangle by hexagons of maximal size (Fig. 5). The residual set is such that $e = \Delta(E) = 1$. It also applies to the packing of a right, isoceles triangle U by regular octogons

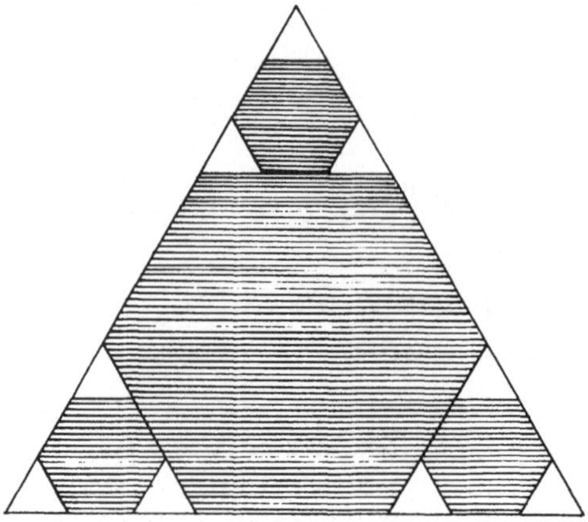

Fig. 5

(Fig. 6): if $\mathscr{C}_1$ is the largest octogon in U, we divide $U-\mathscr{C}_1$ into five copies of U (if the base of U is 1, two of them have base $a=1/(1+\sqrt{2})$, the other three a^2), into each we take the largest octogon, etc. The residual set is such that $e=\Delta(E)=\log 3/\log(1+\sqrt{2})$.

Application 7.4. An example where the $\mathscr{C}_n$ are not disjoint is the following. Take three disks D_1, D_2, D_3 such that D_2 is externally tangent to D_1, D_3 and the center of D_3 belongs to the perimeter of D_1 (Fig. 7). They delimit a curvilinear triangle T_1. The disk D_4 is tangent to D_1, D_3 and its center belongs to the perimeter of D_2. Now three curvilinear triangles $T_{1,1}$, $T_{1,2}$, $T_{1,3}$ are delimited. Perform the same operation in each: such a triangle is bounded by three disks D_1', D_2', D_3' such that D_2' is tangent to D_1', D_3' and the center of D_3' belongs to the perimeter of D_1'; define a new disk D_4' as above. Now assume the process is continued ad infinitum, and take for $(\mathscr{C}_n)$ the family of all these disks. Take for U the inside of $T_1\cup D_1\cup D_2\cup D_3$. Condition (iii) is verified ($N=2$). Theorem 1.1 proves that $e=\Delta(E)$. The value of e has not been calculated. Since E contains no arcs, not even from the perimeters of the $\mathscr{C}_n$, it would be interesting to check if $e>0$ or

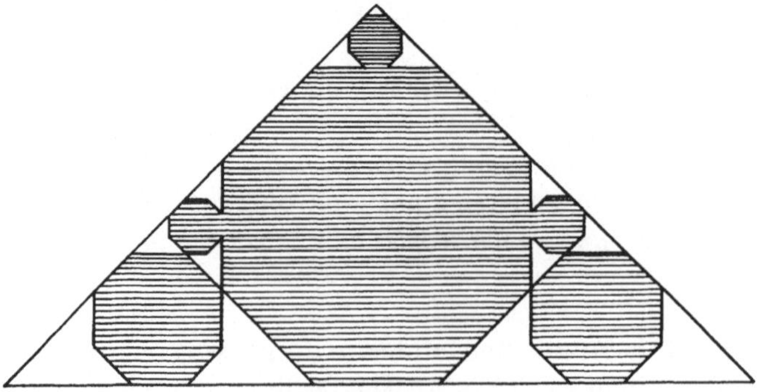

Fig. 6

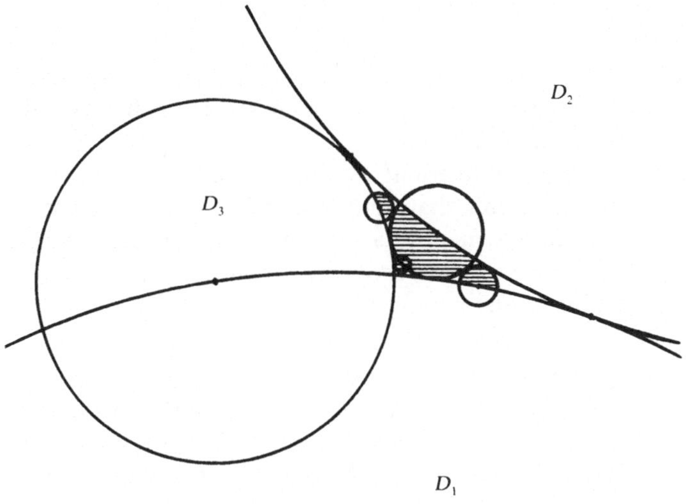

Fig. 7

if $e \leq 1$. E is nonempty since it is the limit of the set of all centers of disks, and this limit contains at least one point.

Application 7.5. It is not necessary to construct packings only by convex sets. Figure 8 is a packing by figures all similar to a regular cross of 12 sides. The residual set is a cartesian product, namely $F \times F$ where F is the triadic Cantor set: $e = \Delta(E) = \log 4/\log 3$. As F, this set E is self-similar. We now see how to generalize such a construction to an important class of self-similar sets.

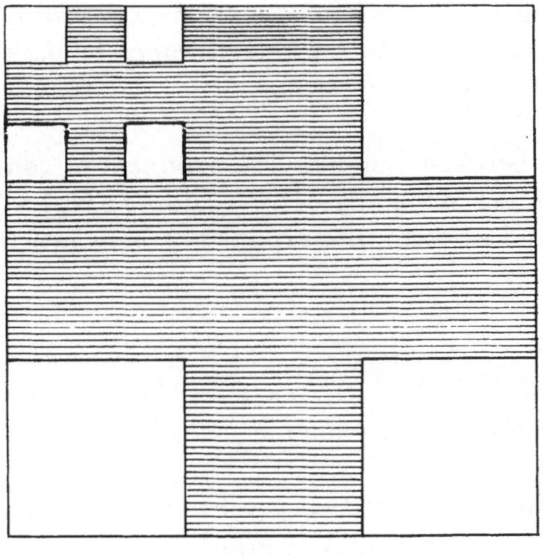

Fig. 8

8. Self-Similar Sets

Definition 8.1. A finite family $\mathcal{F}=\{f_1,\ldots,f_k\}$ of similitudes satisfies the *open set condition* if there exists a nonempty open set U such that $\bigcup_{i=1}^k f_i(U)\subset U$ and $f_i(U)\cap f_j(U)=\emptyset$ if $i\neq j$.

Such a family defines an invariant compact set E called *self-similar*; see [11] for a reference. We can show that, under slightly more restricted conditions, a self-similar set is the complement of a sequence $(\mathscr{C}_n)$ of disjoint open sets, as in Theorem 1.1, which itself possesses property of similarity, that is, for some integer K, every $\mathscr{C}_n$ is similar to one amongst $\mathscr{C}_1, \mathscr{C}_2, \ldots \mathscr{C}_K$.

Proposition 8.2. *Let E be self-similar, included in an open set U as in Definition* 8.1, *such that U is homeomorphic to* S_2 *and has a rectifiable boundary. We can find a disjoint family* $(\mathscr{C}_n)_{n\geq 1}$ *of complementary sets in U, verifying conditions* (i)-(vi) *of Theorem* 1.1, *which for some integer K can be arranged as follows*:

$$\{\mathscr{C}_n\colon n\geq 1\}=\bigcup_{q=1}^{K}\bigcup_{p=1}^{\infty}\{D^{(q)}_{i_0,i_1,\ldots,i_p:i_0=0},\, 1\leq i_j\leq k \text{ for } j\geq 1\},$$

where $D^{(q)}_{i_0,\ldots,i_{p-1},i_p}=f_{i_p}\circ f_{i_{p-1}}\circ\cdots\circ f_{i_1}(\mathscr{C}_q)$, $1\leq q\leq K$.

Proof. All we have to do is to decompose $U-\overline{\bigcup_{i=1}^k f_i(U)}$ into a finite family $\mathscr{C}_1,\ldots,\mathscr{C}_K$ of disjoint open sets homeomorphic to S_2, and denote them $\mathscr{C}_1=D_0^{(1)},\ldots,\mathscr{C}_K=D_0^{(K)}$. Thereafter the other $\mathscr{C}_n$ are defined in every $f_i(U)$ by similarity, and so forth. The integer q being fixed, the family $(D^{(q)}_{i_0,\ldots,i_p})$ is made of sets all similar to $\mathscr{C}_q$. The residual set $E=\bar{U}-\bigcup_1^\infty \mathscr{C}_n$ is the invariant set of the family $\mathcal{F}$ of similitudes. Since $\mathcal{F}$ verifies the open set condition, E has measure 0. Verification of conditions (i)-(vi) are left to the reader. In particular it follows that $e=\Delta(E)$. ∎

Corollary 8.3. *Using the same notation and E as in Proposition* 8.2, *let* $\rho_i=\operatorname{Lip} f_i$ *be the ratio of the similitude* f_i, *and d be determined by the equality* $\sum_{i=1}^k \rho_i^d=1$. *Then* $d=\Delta(E)$.

Proof. Let us fix q between 1 and K. The exponent of convergence of $(\mathscr{C}_n)$ is the same as that of the subfamily $(D^{(q)}_{i_0,\ldots,i_p})$ for $p\in\mathbf{N}, 1\leq i_j\leq k$. Since $D^{(q)}_{i_0}=\mathscr{C}_q$, $D^{(q)}_{i_0,i_1}=f_{i_1}(\mathscr{C}_q)$, and so forth, the exponent e is in fact the critical index of the series

$$\sum_{p=1}^{\infty}\left(\sum_{i_1=1}^{k}\cdots\sum_{i_p=1}^{k}(\rho_{i_1}\cdots\rho_{i_p})^\alpha\right)=\sum_{p=1}^{\infty}(\rho_1^\alpha+\cdots+\rho_k^\alpha)^p;$$

therefore e is the solution of the equation $\rho_1^\alpha+\cdots+\rho_k^\alpha=1$. ∎

Remark 8.4. We know [14] that d is also the Hausdorff dimension of E. As to $\Delta(E)$, there are of course other ways to prove Corollary 8.3. This proof implies in particular that $\sum_1^\infty c_n^d$ is infinite.

Fig. 9

Example 8.5. The self-similar set E of Fig. 9 is the same as the one in Fig. 5, only the starting domain U is different. Indeed, the invariant set E is unique [11], depending only on the three similitudes, and not on U. The shaded area indicates the union of four complementary sets $\mathscr{C}_1$-$\mathscr{C}_4$ (here $K=4$). The other $\mathscr{C}_n$ are similar to the first four.

9. Weakening the Conditions of Theorem 1.1

We still assume that condition (iv) is unchanged. All the other conditions deal with characteristic ratios ($c_n^2/|\mathscr{C}_n|$, l_n/c_n), etc., assumed to be bounded. As very often in fractal theory such ratios may be replaced by ratios of logarithms, as described in Theorem 9.1. This possible weakening is due to the fact that $\Delta(E)$, as well as the Hausdorff dimension, is a critical index in a family of orders of growth, namely those of the power functions t^α, $\alpha \geq 0$. This family is quite small, compared with other scales of functions. A further theory could refine these indices by using larger scales, and then such a generalization as the following theorem may no longer be true.

Theorem 9.1. *Consider an infinite collection $U=\mathscr{C}_0, \mathscr{C}_1, \mathscr{C}_2, \ldots$ of sets homeomorphic to the open disk S_2, such that $\mathscr{C}_n \subset U$ if $n \geq 1$. The curve $\partial\mathscr{C}_n$ is rectifiable for $n \geq 0$, of length $L(\partial\mathscr{C}_n)=l_n$. We denote by M_n the minimum number of families of disjoint sets in which $\{\mathscr{C}_i\}_{1\leq i\leq n}$ can be partitioned, by U_n the union of all $\mathscr{C}_i$ such that $c_i \geq c_n$, and by K_n the number of connected components in $U-\bar{U}_n$. Take for residual set E any nonvoid Borel set such that*

$$U-\bigcup_{i=1}^{\infty} \bar{\mathscr{C}}_n \subset E \subset \bar{U}-\bigcup_{i=1}^{\infty} \mathscr{C}_n.$$

Let $c_n = \operatorname{diam} \mathscr{C}_n$. *We consider the following six properties*:

(i) $\limsup \log|\mathscr{C}_n|/\log c_n \leq 2$;
(ii) $\limsup \log c_n/\log \operatorname{dist}(\mathscr{C}_n, E) \leq 1$;
(iii) $\limsup \log M_n/\log n \leq 1$;
(iv) $|\bigcup \mathscr{C}_n| = |U|$;
(v) $\limsup \log c_n/\log l_n \leq 1$;
(vi) $\limsup \log K_n/\log n \leq 1$.

Then the conclusion of Theorem 1.1 *holds.*

We omit the details of the proof, which uses the same essential arguments as Theorem 1.1. They only require rather more analytic computations.

10. Generalizing to $\mathbf{R}^p$, $p \geq 3$

The definition of $\Delta(E)$ in $\mathbf{R}^p$ is

$$\Delta(E) = \inf\{\alpha: r^{\alpha-p}|E(r)| \to 0\}, \tag{11}$$

where $|\,|$ is the p-dimensional measure. A good part of Theorem 1.1 is still valid in $\mathbf{R}^p$. We have to assume that U and $\mathscr{C}_n$, $n \geq 1$, are homeomorphic to the open ball S_p, and that their boundary is rectifiable, that is, in terms of Hausdorff measure Λ^{p-1}, for all $n \geq 0$, $0 < \Lambda^{p-1}(\partial\mathscr{C}_n) < +\infty$. We denote $l_n = \Lambda^{p-1}(\partial\mathscr{C}_n)$. It is then straightforward to prove that

$$e \leq \Delta(E) \leq \max(e, p-1)$$

if the following five conditions are verified: (i) $c_n^p = \mathcal{O}(|\mathscr{C}_n|)$, (ii)-(v) as in Theorem 1.1. In particular, a generalization of Lemma 4.3 presents no difficulty.

Problems arise for part B2 of Theorem 1.1, that is, in the case where $\sum c_n^{p-1} < +\infty$. Certainly, condition (vi) should be implemented. It would be interesting to study at least the case of convex bodies. Our conjecture is as follows:

Conjecture 10.1. *Assume that U is, in* $\mathbf{R}^p$, *the complement of a finite union of closed, convex domains, and that the* $\mathscr{C}_n$, $n \geq 1$, *are convex open sets in U. We define E, and conditions* (i)-(v), *as in Theorem* 1.1, *with the suitable transformations from dimension* 2 *to dimension p. Then* $e = \Delta(E)$.

A proof seems to be connected with the geometry of *starlike sets.* We can state a very special case:

Proposition 10.2. *Let U be an open set in* $\mathbf{R}^p$, *homeomorphic to* S_p, *with a rectifiable boundary, and the* $\mathscr{C}_n$, $n \geq 1$, *be disjoint open cubes with parallel sides, included in U. If* $\operatorname{dist}(\mathscr{C}_n, E) = \mathcal{O}(c_n)$, $\bar{E} \subset U$, *and* $|E| = 0$, *then* $e = \Delta(E)$.

For a proof, see [16, *Etude dimensionnelle d'un compact par le pavage du complémentaire*]. Here again Propositions 2.1 and 2.2 are very useful, considering such indices as $e^*(c_n; 0, 0, p-q)$ and $e^{**}(c_n; 0, 0, p-q)$ following the convergence of the series $\sum c_n^q$, $1 \leq q \leq p-1$.

Particular Case 10.3. We show that every compact set E of zero measure in $\mathbf{R}^p$ is the complement of a family $(\mathscr{C}_n)$ of cubes as in Proposition 10.2. Consider a net of dyadic open cubes, of the form $\prod_{i=1}^{p}]k_i 2^{-n}, (k_i+1)2^{-n}[$ where $k_1, \ldots, k_p \in \mathbf{Z}$, $n \in \mathbf{N}$. Two such cubes are either disjoint or imbedded. Take for U the smallest dyadic cube containing E, for $\mathscr{C}_1$ the largest (or one of the largest) dyadic cube in $U - E$, such that $\text{dist}(\mathscr{C}_1, E) \geq c_1$, for $\mathscr{C}_2$ the largest dyadic cube in $U - \mathscr{C}_1 - E$ such that $\text{dist}(\mathscr{C}_2, E) \geq c_2$, etc. The family $(\mathscr{C}_n)$ thus defined by induction is called a *Whitney family* for E, where $E = U - \bigcup_1^\infty \bar{\mathscr{C}}_n$. Since $\text{dist}(\mathscr{C}_n, E) \leq 4c_n$, Proposition 10.2 holds and $\Delta(E) = e$. Actually, a much shorter, direct proof is available for Whitney families in [16]: the main argument is that, for r small enough, and n such that $c_{n+1} < r \leq c_n$, the p-measure of the set $E(r) - \bigcup_{n+1}^\infty \mathscr{C}_i$ is zero.

It is interesting to note that a Whitney family of cubes provides, for any compact set E, an example of a sequence $(\mathscr{C}_n)$ verifying all conditions of Theorem 1.1 (in $\mathbf{R}^2$) or of Conjecture 10.1 (in $\mathbf{R}^p$).

11. Questions

The following are two, partly open, problems about disjoint disk packings.

Question 11.1. Given any $\alpha > \frac{1}{2}$, is it possible to arrange in a square U a disjoint family of disks $(\mathscr{C}_n)$ such that $e = \alpha$ and $\bigcup_1^\infty \mathscr{C}_n$ is dense in U?

Question 11.2. In the same context, can $|U - \bigcup_1^\infty \mathscr{C}_n|$ be zero?

For both questions we already know that the answer is yes for $\alpha_0 \leq \alpha \leq 2$ [19], [4] where α_0 is the apollonian tacking constant. It is proved that if the residual set E has measure zero, then $\dim(E) \geq 1.03$ where dim is the Hausdorff dimension. Since the inequality $\dim E \leq \Delta(E)$ is always true, we deduce that the answer to 11.2 is no for $\frac{1}{2} < \alpha < 1.03$, and conjecture that the same holds for $\alpha < \alpha_0$.

Other problems look for the smallest domain, of prescribed shape, containing a given family of disks. For example:

Question 11.3. Find the smallest s for which an open domain U, homeomorphic to S_2, of area s, contains disjoint disks of diameters $c_n = n^{-\alpha}$, $\alpha > \frac{1}{2}$. If such a U exists for the minimal value of s, is ∂U rectifiable, and the disks everywhere dense in U?

For related problems see [7].

Acknowledgments. We wish to thank S. J. Taylor and D. W. Boyd for useful discussions on the subject, while preparing this paper.

References

1. A. S. Besicovitch, S. J. Taylor (1954): *On the complementary intervals of a linear closed set of zero Lebesgue measure.* J. London Math. Soc., **29**:449–459.

2. E. Borel (1948): *Sur l'addition vectorielle des ensembles de mesure nulle.* C. R. Acad. Sci. Paris, **227**:103–105.
3. G. Bouligand (1929): *Sur la notion d'ordre de mesure d'un ensemble plan.* Bull. Sci. Math., **53**:185–192.
4. D. W. Boyd (1972): *Disk packings which have non-extreme exponents.* Canad. Math. Bull., **15**:341–344.
5. D. W. Boyd (1973): *Improved bound for the disk packing constant.* Aequationes Math., **9**:99–106.
6. D. W. Boyd (1973): *The residual set dimension of the apollonian packing.* Mathematika, **20**:170–174.
7. D. W. Boyd (1980): *Problem* 276. Canad. Math. bull., **23**:251-253.
8. Y. Dupain, M. Mendès-France, C. Tricot (1983): *Dimensions de spirales.* Bull. Soc. Math. France, **111**:193-201.
9. C. Grebogi, S. McDonald, E. Ott, J. A. Yorke (1985): *Exterior dimension of fat fractals.* Phys. Lett. A, **110**:1-4.
10. J. Hawkes (1974): *Hausdorff measure, entropy and the independence of small sets.* Proc. London Math. Soc., **28**:700-724.
11. J. E. Hutchinson (1981): *Fractals and self-similarity.* Indiana Univ. Math. J., **30**:713–747.
12. B. Mandelbrot (1982): The Fractal Geometry of Nature. San Francisco: Freeman.
13. Z. A. Melzak (1966): *Infinite packing of disks.* Canad. J. Math., **18**:838-852.
14. P. A. P. Moran (1946): *Additive functions of intervals and Hausdorff measure.* Math. Proc. Cambridge Philos. Soc., **42**:15–23.
15. C. Tricot (1981): *Douze définitions de la densité logarithmique.* C. R. Acad. Sci. Paris, **293**:549-552.
16. C. Tricot (1983): Mesures et dimensions. Thèse de doctorat, Orsay.
17. C. Tricot (1984): *A new proof for the residual set dimension of the apollonian packing.* Math. Proc. Cambridge Philos. Soc., **96**:413-442.
18. C. Tricot (1986): *The geometry of the complement of a fractal set.* Phys. Lett. A, **114**:8–9, 430-434.
19. C. Tricot (1987): *Dimensions aux bords d'un ouvert.* Ann. Sci. Math. Québec, **11**:205-235.
20. J. B. Wilker (1965): Open Disk Packings of a Disk. Ph.D. Thesis, Toronto.
21. J. B. Wilker (1973): *The interval of disk packing exponents.* Proc. Amer. Math. Soc., **41**:255-260.
22. J. B. Wilker (1977): *Sizing up a solid packing.* Period. Math. Hungar., **8**:117–134.
23. U. Zähle (1984): *Random fractals generated by random cutouts.* Math. Nachr., **116**:27-52.

C. Tricot
Ecole Polytechnique de Montréal
Département de mathématiques appliquées
C.P. 6079, Succursale "A"
Montréal (Québec)
Canada H3C 3A7

Constr. Approx. (1989) 5: 137–150

CONSTRUCTIVE APPROXIMATION

Interpolating and Orthogonal Polynomials on Fractals

Hans Wallin

Abstract. A special class of closed subsets F of $\mathbf{R}^n$, referred to as sets preserving Markov's inequality, are considered. Typically, F may be a fractal such as the Cantor set or von Koch's curve, but F may also be a closed Lipschitz domain. We investigate interpolation to smooth functions on F where the points of interpolation belong to F. We also consider orthogonal polynomials on $F \cap B$, where B is a ball with center in F, and their relation to spaces of smooth functions on F.

Introduction. Analysis on Fractals

Traditionally, the functions studied in analysis are assumed to be defined everywhere in n-dimensional space $\mathbf{R}^n$ or on a subset of $\mathbf{R}^n$ with a rather regular geometry. One reason for this is, of course, that this gives a simple situation and a rich theory where a lot of different methods may be used. Another reason is the fundamental influence from the theory of differential equations where, typically, you study functions defined on a nice domain. More recently, however, various phenomena in the sciences have been described using very irregular sets. These are sets of fractional Hausdorff dimension (see Section 1.1). B. B. Mandelbrot [9] and others have used these sets in applications in a systematic way during the latest decades and the sets are today widely known as fractals. For example, fractals are used to describe nonlinear dynamical systems such as turbulence in fluids, Brownian motion of particles, and geographical coastlines. Formally, following Mandelbrot [9] we call a set $F \subset \mathbf{R}^n$ a fractal if F has fractional Hausdorff dimension or if F has integral Hausdorff dimension and is irregular in the sense that the Hausdorff dimension of F is strictly larger than the topological dimension of F. Mathematically, the study of fractals goes back to Cantor's set theory. The geometric theory of fractals is rich and was pioneered by Besicovitch who started his investigation in this area around 1920 (see [3]). Fractals have also been used in harmonic analysis, for example in a monograph from 1963 by Kahane–Salem.

Date received: March 4, 1987. Communicated by Michael F. Barnsley.
AMS classification: Primary 41A63, 46E35; Secondary 41A05, 41A10.
Key words and phrases: Fractals, Polynomial interpolation, Orthogonal polynomials, Function spaces, Local approximation, Several variables, Markov's inequality.

In recent years a systematic study of spaces of smooth functions on fractals and Hardy spaces on fractals has started (see [7] and, for Hardy spaces, [8]) although some such material can be traced further back, for instance, to Whitney's work in the 1930s. There is a unified way of describing spaces of smooth functions defined on $F \subset \mathbf{R}^n$ by using local approximation by polynomials on F intersected by squares or balls B. This is a kind of spline-type approximation where the degree of approximation on $B \cap F$ and the degree of the polynomials essentially equal the order of smoothness of the function which is approximated. This description becomes particularly simple (see Section 2) if F has a certain property related to Markov's inequality for polynomials—F preserves Markov's inequality (see Section 1.2). In Section 3 we show how orthogonal polynomials on fractals F can be defined and used to describe these smooth function spaces on F in a more constructive way. In Section 4 we investigate approximation by interpolation by algebraic polynomials in several variables where the interpolation points are chosen from a given fractal. This section is based on results by Ciarlet and Raviart [1]. In Sections 3 and 4 we assume that F preserves Markov's inequality.

It should be stated explicitly that even if the sets which we primarily have in mind in this paper are fractals preserving Markov's inequality, the results are true for any sets F preserving Markov's inequality. The point we want to stress, however, is that the methods are applicable not only to certain sets with a nice geometry, like a Lipschitz domain or $\mathbf{R}^n$ itself, but also to certain very irregular sets.

Notation. $F \subset \mathbf{R}^n$ is a closed, nonempty set (usually preserving Markov's inequality) and $B(x, r)$ is the closed n-dimensional ball with center $x \in F$ and radius $r \leq 1$. μ is a positive, nontrivial Borel measure finite on bounded sets, with support supp μ. $\mathcal{P}_k$ is the set of all polynomials in n real variables of total degree at most k. D_j stands for the partial derivative corresponding to a multi-index $j = (j_1, \ldots, j_n)$ with length $|j| = j_1 + \cdots + j_n$. D^k stands for the kth Fréchet derivative (Section 4). $C^k(F)$, $\mathrm{Lip}(\alpha, F)$, and $\Lambda_\alpha(F)$ are spaces of smooth functions on F (Section 2). c, c_1, c_2 denote different positive (finite) constants.

1. Sets Preserving Markov's Inequality

1.1. Hausdorff Measures, d-Measures, and the Doubling Condition

We first recall the definition of the *d-dimensional Hausdorff measure of E*, $m_d(E)$, where d is a positive number and $E \subset \mathbf{R}^n$.

$$m_d(E) = \lim_{\varepsilon \downarrow 0} m_d^{(\varepsilon)}(E),$$

where, for some positive constant $\alpha(d)$,

$$m_d^{(\varepsilon)}(E) = \alpha(d) \inf\left\{ \sum_i (\operatorname{diam} E_i)^d : \bigcup_1^\infty E_i \supset E, \operatorname{diam} E_i \leq \varepsilon \right\}.$$

Here diam E_i is the diameter of E_i. We choose $\alpha(d)$ so that $m_n(E)$ coincides with the n-dimensional outer Lebesgue measure of E. The d-dimensional Hausdorff measure is an outer metric measure and the class of sets measurable m_d contains the Borel sets in $\mathbf{R}^n$. The *Hausdorff dimension of E*, $\dim(E)$, is the infinum of the set of $d>0$ such that $m_d(E)=0$. It is easy to see that $0\leq\dim(E)\leq n$. We refer to [3] for more details.

Next, we define the concepts of d-measure and d-set. The set F is a *d-set* $(0<d\leq n)$ if there exists a μ with $\operatorname{supp}\mu=F$ such that, for some positive constants $c_1=c_1(F)$ and $c_2=c_2(F)$,

$$c_1r^d\leq\mu(B(x,r))\leq c_2r^d \qquad \text{for} \quad x\in F \text{ and } 0<r\leq 1.$$

Such a μ is called a *d-measure on F*. We refer to Chapter II of [7] for these concepts as well as for the following properties. If F is a d-set then $m_d|F$, the restriction to F of the d-dimensional Hausdorff measure, is a d-measure on F. Also, any d-measure μ on F is equivalent to $m_d|F$ in the sense that, for some constants c' and c'', $c'\mu\leq m_d|F\leq c''\mu$. Finally, if F is a d-set, then $\dim(F)=d$ and $\dim(F\cap B(x,r))=d$ for all $x\in F$ and $r>0$. $\mathbf{R}^n$ itself and a closed domain in $\mathbf{R}^n$ with boundary locally in Lip 1 are examples of d-sets with $d=n$ where the Lebesgue measure gives the d-measure. We also give two examples of fractals F which are d-sets and where it is easy to construct a d-measure on F and, consequently, to determine $\dim(F)$.

Example 1. *The Cantor set.* Let $F\subset\mathbf{R}^1$ $(n=1)$ be the ordinary Cantor set $F=\bigcap F_i$ where $F_0=[0,1]$ and F_i is the union of 2^i closed intervals of length 3^{-i} obtained by removing the middle thirds of the intervals of F_{i-1}. If μ_i with support F_i is the absolutely continuous probability measure on F_i with constant density on F_i, μ_i converges weakly to a measure μ supported by F and it is easy to check that μ is a d-measure on F for $d=\log 2/\log 3$. Hence, F is a d-set for this d and $\dim(F)=\log 2/\log 3$.

Example 2. *Von Koch's snowflake curve.* Let F_0 be the boundary of an equilateral triangle with sides of unit length and replace the middle third of each side of F_0 by the two other sides of an equilateral triangle with sides of length 3^{-1} so that the new sides are located outside the original triangle. This gives F_1 consisting of 3×4 sides of length 3^{-1}. Now we repeat this process to each of the sides of F_1 and get F_2 consisting of 3×4^2 sides of length 3^{-2}, and so on. After i steps we get F_i consisting of 3×4^i sides of length 3^{-i}. As i tends to infinity, F_i converges in a natural way to a set F, von Koch's curve; the convergence is technically described as convergence in the Hausdorff metric (see p. 37 of [3]). In the same way as in Example 1 it is easy to check that F is a d-set, this time with $d=\log 4/\log 3$, and hence $\dim(F)=\log 4/\log 3$.

If F is a d-set and μ a d-measure on F, then μ satisfies the *doubling condition*: for some constant $c=c(F)$ (the doubling constant)

$$\mu(B(x,2r))\leq c\mu(B(x,r)) \qquad \text{for} \quad x\in F \text{ and } 0<r\leq 1.$$

It has been proved by Volberg and Konyagin [12] that any closed subset of $\mathbf{R}^n$ is the support of a μ satisfying the doubling condition.

1.2. Sets Preserving Markov's Inequality

We are interested in sets F satisfying the condition in the following definition. For this definition as well as for the properties stated below we refer to Chapter II of [7] and, for Property 5, to [16].

Definition. $F \subset \mathbf{R}^n$ *preserves Markov's inequality* if for every positive integer k there exists a constant $c = c(n, k, F)$ such that, for all polynomials $P \in \mathcal{P}_k$ and all balls $B = B(x_0, r)$, $x_0 \in F$, $0 < r \leq 1$, we have

$$\max_{F \cap B} |\nabla P| \leq \frac{c}{r} \max_{F \cap B} |P|. \tag{1.1}$$

We refer to (1.1) as *Markov's inequality* on F and note that for $F = \mathbf{R}^n$ it is the ordinary Markov inequality in $\mathbf{R}^n$. We are not concerned here with finding the best or even a good constant c but are just interested in using (1.1) as a condition on F. The following properties hold.

Property 1. F preserves Markov's inequality iff

$$\max_{B} |P| \leq c \max_{F \cap B} |P|, \qquad c = c(n, k, F), \tag{1.2}$$

for all k, all $P \in \mathcal{P}_k$, and all $B = B(x_0, r)$, $x_0 \in F$, $0 < r \leq 1$.

For the only-if-part we refer to p. 35 of [7], and the if-part follows for example from the following estimates where we first use a trivial estimate, then Markov's inequality in $\mathbf{R}^n$, and finally (1.2):

$$\max_{F \cap B} |\nabla P| \leq \max_{B} |\nabla P| \leq \left| \frac{c_1}{r} \right| \max_{B} |P| \leq \left| \frac{c_2}{r} \right| \max_{F \cap B} |P|.$$

Property 2. If F preserves Markov's inequality and μ with supp $\mu = F$ satisfies the doubling condition, then, for $1 \leq q < \infty$,

$$\max_{F \cap B} |P| \leq c \left\{ \frac{1}{\mu(B)} \int_B |P|^q \, d\mu \right\}^{1/q} \tag{1.3}$$

for all k, all $P \in \mathcal{P}_k$, and all $B = B(x_0, r)$, $x_0 \in F$, $0 < r \leq 1$. The constant c in (1.3) depends only on n, k, F and the doubling constant of μ. Properties 1 and 2 are used in Section 3.

Property 3 (*Geometric Characterization*). F preserves Markov's inequality iff, for some $\varepsilon > 0$, none of the sets $B \cap F$, where $B = B(x_0, r)$, $x_0 \in F$, $0 < r \leq 1$, is contained in any band of type

$$\{x \in \mathbf{R}^n : |b \cdot (x - x_0)| < \varepsilon r\} \qquad \text{where} \quad b \in \mathbf{R}^n \text{ and } |b| = 1.$$

The property means that F is locally never too flat but that F in a way extends in all n dimensions. This property is used to prove

Property 4. If $F \subset \mathbf{R}^n$ is a d-set with $d > n - 1$, then F preserves Markov's inequality.

Property 4 provides us with a large collection of sets F preserving Markov's inequality, for instance, the sets in Examples 1 and 2. Other examples are $\mathbf{R}^n$ itself and the closure of a domain with boundary locally in Lip 1. Property 3 has been used by P. Wingren [16] to give the following characterization of sets preserving Markov's inequality.

Property 5. F preserves Markov's inequality iff there exists a constant $c^*>0$ so that, for every $B=B(x_0, r)$ where $x_0 \in F$ and $0<r\leq 1$, there are $n+1$ affinely independent points $a_i \in F \cap B$, $i=1,\ldots,n+1$, such that the n-dimensional ball inscribed in the convex hull of $a_1, a_2, \ldots, a_{n+1}$ has radius not less than $c^* r$.

In Section 4 we use Property 5 to study polynomial interpolation.

2. Function Spaces on F

2.1. A General F

We introduce three spaces of smooth functions on F. We first describe them on $\mathbf{R}^n$. Let k be a nonnegative integer and $C^k(\mathbf{R}^n)$ be the space of continuous functions in $\mathbf{R}^n$ having continuous partial derivatives up to order k. In $C^k(\mathbf{R}^n)$ we introduce the usual topology defined by the family of seminorms

$$|f; C^k(K)| = \max_{|j|\leq k} \max_{K} |D_j f|, \tag{2.1}$$

where $K \subset \mathbf{R}^n$ are compact. If $k<\alpha\leq k+1$, $\mathrm{Lip}(\alpha, \mathbf{R}^n)$, the Lipschitz space of order α corresponding to first differences, is the space of functions $f \in C^k(\mathbf{R}^n)$ such that

$$|D_j f(x)| \leq M \qquad \text{for} \quad |j|\leq k \text{ and } x \in \mathbf{R}^n$$

and

$$|\Delta_h D_j f(x)| \leq M|h|^{\alpha-k} \qquad \text{for} \quad |j|=k \text{ and } x, h \in \mathbf{R}^n.$$

Here $\Delta_h g(x) = g(x+h)-g(x)$ and the norm $\|f; \mathrm{Lip}(\alpha, \mathbf{R}^n)\|$ is the infimum of all possible constants M in these two inequalities. We get the space $\Lambda_\alpha(\mathbf{R}^n)$, the Lipschitz space of order α corresponding to second differences, if the first difference Δ_h is changed to the second difference Δ_h^2. The two Lipschitz spaces differ only if $\alpha=k+1$ and we do not get any new spaces by using higher-order differences.

For any closed $F \subset \mathbf{R}^n$ it is possible to describe the corresponding function spaces on F, $C^k(F)$, $\mathrm{Lip}(\alpha, F)$, and $\Lambda_\alpha(F)$, so that these are trace spaces to F of the corresponding spaces in $\mathbf{R}^n$. More precisely, if $S(\mathbf{R}^n)$ denotes any of $C^k(\mathbf{R}^n)$, $\mathrm{Lip}(\alpha, \mathbf{R}^n)$, and $\Delta_\alpha(\mathbf{R}^n)$, and $S(F)$ the corresponding space on F, then $S(F) = S(\mathbf{R}^n)|F$ (trace theorem) where $S(\mathbf{R}^n)|F$ is the trace space defined by

$$S(\mathbf{R}^n)|F = \{\{f^{(j)}\}_{|j|\leq k} : f^{(j)} = (D_j f)|F, \qquad \text{for} \quad |j|\leq k,\ f \in S(\mathbf{R}^n)\}.$$

Here $(D_j f)|F$ denotes the pointwise restriction to F of $D_j f$. For $C^k(\mathbf{R}^n)$ and $\mathrm{Lip}(\alpha, \mathbf{R}^n)$ this goes back to Whitney [15] (Whitney's extension theorem) and for $\Lambda_\alpha(\mathbf{R}^n)$ to A. Jonsson and the author [6]; see also [7], [11], and [10]. The elements of $S(F)$ are, consequently, families of functions $\{f^{(j)}\}_{|j|\le k}$ defined on F where $f^{(j)}$ for $|j|>0$ stand for formal partial derivatives of $f^{(0)}$. For instance, by Whitney's method the space $C^k(F)$ is described as follows where, for simplicity, we assume that F is compact. The family $\{f^{(j)}\}_{|j|\le k}\in C^k(F)$ if, firstly, there exists a constant M such that if

$$R_j(x,y)=f^{(j)}(x)-\sum_{|j+l|\le k}\frac{f^{(j+l)}(y)}{l!}(x-y)^l$$

then, for all $x, y\in F$ and $|j|\le k$,

$$|f^{(j)}(x)|\le M \quad\text{and}\quad |R_j(x,y)|\le M|x-y|^{k-|j|}, \tag{2.2}$$

and, secondly, for $|j|\le k$,

$$R_j(x,y)=o(|x-y|^{k-|j|}) \qquad \text{if}\quad x,y\in F,\ \text{when } |x-y|\to 0.$$

The norm $\|\{f^{(j)}\}; C^k(F)\|$ is the infimum of all constants M such that (2.2) holds. Then, Whitney's extension theorem (see [11]) states that there exists a linear extension operator $E: C^k(F)\to C^k(\mathbf{R}^n)$ such that the extension Ef of $f=\{f^{(j)}\}_{|j|\le k}$ has $f^{(j)}$, $|j|\le k$, as partial derivatives on F. Furthermore [11, p. 75], [4], if $K\supset F$ is compact then the extension Ef of $f=\{f^{(j)}\}$ satisfies, using the notation of (2.1),

$$|Ef; C^k(K)|\le c\|f; C^k(F)\|, \tag{2.3}$$

with a constant c depending only on n, k, and $\sup\{d(x,F): x\in K\}$ where $d(x,F)$ is the distance from x to F. This is used in Section 4.

2.2. *F Preserves Markov's Inequality*

Now, in the rest of this paper, assume that F preserves Markov's inequality and let $S(\mathbf{R}^n)|F=S(F)$ be any of the three spaces introduced above and take an element $\{f^{(j)}\}_{|j|\le k}\in S(\mathbf{R}^n)|F$. Then we claim that $f^{(j)}$, $0<|j|\le k$, are uniquely determined by $f^{(0)}$ in the sense that if $\{g^{(j)}\}_{|j|\le k}\in S(\mathbf{R}^n)|F$ is such that $g^{(0)}=f^{(0)}$, then $g^{(j)}=f^{(j)}$ for $0<|j|\le k$. In fact, $f^{(j)}=(D_jf)|F$ for some $f\in C^k(\mathbf{R}^n)$ and if $f=0$ on F then $D_jf=0$ on F for $|j|\le k$ [7, p. 41], [4]. Hence, $f^{(0)}=0$ on F implies $f^{(j)}=0$ on F, for $0<|j|\le k$, and the claim follows by linearity. Since $\{f^{(j)}\}_{|j|\le k}\in S(\mathbf{R}^n)|F=S(F)$ is uniquely determined by $f^{(0)}$ we identify $\{f^{(j)}\}$ with $f^{(0)}$ and speak of $f^{(0)}\in S(F)$ and of the partial derivatives $D_jf^{(0)}=f^{(j)}$, $0<|j|\le k$, of $f^{(0)}$ on F.

Above we described $C^k(F)$ by Whitney's method. Lip(α, F) may be described analogously [10], [7]. Another way to describe the spaces $S(F)=S(\mathbf{R}^n)|F$ is by local polynomial approximation. For $\Lambda_\alpha(F)$ this is done in the following definition where $[\alpha]$ denotes the integer part of α (we refer to Chapter III of [7], for details; see also [14]).

Definition 2.1. $f\in\Lambda_\alpha(F)$, $0\le k<\alpha\le k+1$, k integer, if f is defined on F and, for some constant M, $|f|\le M$ on F and, for every ball $B=B(x_0,r)$, $x_0\in F$, $0<r\le 1$, there exists a polynomial $P_B\in\mathcal{P}_{[\alpha]}$ such that

$$|f(x)-P_B(x)|\le Mr^\alpha \qquad \text{for}\quad x\in B\cap F.$$

The norm of f in $\Lambda_\alpha(F)$ is defined as the infimum of all possible constants M.

The space Lip(α, F) may be defined as in Definition 2.1 with the only difference that we require $P_B \in \mathcal{P}_k$ (see Chapter III of [7]). Also $C^k(F)$ may be defined using local polynomial approximation [17]. It may be remarked that Besov spaces on F may be defined and described by local polynomial approximation in $L^p(\mu)$-norm where μ is a d-measure on F [7, Chapter V], [14]. We shall, however, not treat Besov spaces here.

3. Orthogonal Polynomials on F

Let $F \subset \mathbf{R}^n$ preserve Markov's inequality and let μ be a measure with support F satisfying the doubling condition (see Section 1.1). Let us consider a fixed ball $B = B(x_0, r)$, $x_0 \in F$, $0 < r \le 1$, and an integer $k \ge 0$. Let $L^2(\mu, B)$ be the L^2 space with inner product

$$(f, g) = \int_B f\bar{g}\, d\mu.$$

We put $g_j(x) = (x - x_0)^j$ and want to orthonormalize the sequence $\{g_j\}_{|j| \le k}$ in $L^2(\mu, B)$ by the Gram-Schmidt process. We first observe that this sequence is linearly independent in $L^2(\mu, B)$. In fact, $\sum c_j g_j = 0$ in $L^2(\mu, B)$ implies, by (1.3) and (1.2), $\sum c_j g_j = 0$ for all $x \in B$ and, hence, $c_j = 0$ for all j. Now, let $\{\varphi_j\}_{|j| \le k}$ be the sequence we obtain by orthonormalization of $\{g_j\}_{|j| \le k}$. This gives us a projection

$$\Pi_B : L^1(\mu, B) \to \mathcal{P}_k$$

defined by

$$\Pi_B f = \sum_{|j| \le k} (f, \varphi_j)\varphi_j. \tag{3.1}$$

We shall need the following lemma.

Lemma 3.1.

$$\max_B |\Pi_B f| \le \frac{c}{\mu(B)} \int_B |f|\, d\mu,$$

for all $f \in L^1(\mu, B)$, *where* $c = c(n, k, F)$.

Proof. Using (1.2), (1.3), and the fact that the $L^2(\mu, B)$ norm of φ_j is 1, we get

$$\max_B |\varphi_j| \le c_1 \max_{B \cap F} |\varphi_j| \le c_2 \left\{ \frac{1}{\mu(B)} \int_B |\varphi_j|^2\, d\mu \right\}^{1/2} = c\mu(B)^{-1/2}.$$

By combining this with (3.1) we get the lemma. ■

The lemma gives

$$\max_B |\Pi_B f| \le c\|f\|_{L^\infty(\mu, B)}, \qquad f \in L^\infty(\mu, B). \tag{3.2}$$

Now we can prove

Proposition 3.2. $\|f - \Pi_B f\|_{L^\infty(\mu, B)} \le c \inf_{P \in \mathcal{P}_k} \|f - P\|_{L^\infty(\mu, B)}$, *for* $f \in L^\infty(\mu, B)$, *where* $c = c(n, k, F)$.

Proof. By the triangle inequality we get for any $P \in \mathcal{P}_k$

$$\|f - \Pi_B f\|_{L^\infty(\mu,B)} \le \|f - P\|_{L^\infty(\mu,B)} + \|\Pi_B f - P\|_{L^\infty(\mu,B)}.$$

Since Π_B is linear and a projection on $\mathcal{P}_k$ we conclude by means of (3.2)

$$\|\Pi_B f - P\|_{L^\infty(\mu,B)} = \|\Pi_B(f - P)\|_{L^\infty(\mu,B)} \le c\|f - P\|_{L^\infty(\mu,B)}$$

and together with the first estimate this gives the proposition. ■

We now state two applications which we formulate for the space $\Lambda_\alpha(F)$ (but similar applications hold for the other spaces).

Application 1. The space $\Lambda_\alpha(F)$ can be defined using the projection operator Π_B. In fact, by using Proposition 3.2 with k replaced by $[\alpha]$ we see that there is a canonical, constructive way to choose the polynomial P_B in Definition 2.1, namely $P_B = \Pi_B f$.

Application 2. If we combine Application 1 with a Whitney type extension method (see Chapter III of [7]) we get the following result: there exists a continuous linear extension operator

$$\Lambda_\alpha(F) \to \Lambda_\alpha(\mathbf{R}^n).$$

Discussion. Orthogonal polynomials on sets preserving Markov's inequality were first used in [8]. The results discussed in this section were proved independently by A. Jonsson [5]. Concerning Application 2 for the space Lip(α, F), see [13]. P. Wingren [16] has proved that the polynomials P_B in Definition 2.1 of the space $\Lambda_\alpha(F)$ may be chosen as interpolating polynomials where the points of interpolation depend only on F and B and not on f. DeVore and Sharpley [2] have studied spaces of smooth functions defined on $\mathbf{R}^n$ or on a Lipschitz domain of $\mathbf{R}^n$ using maximal functions. The approach in this section generalizes some of their basic calculations to sets F preserving Markov's inequality.

4. Interpolation on F

Some general error estimates for Lagrange and Hermite interpolation in $\mathbf{R}^n$ were published by Ciarlet and Raviart [1] in 1972. We use the following notation: let $\Sigma \subset \mathbf{R}^n$ be the set of interpolation points,

$$K = \operatorname{con}(\Sigma) \text{ the closed convex hull of } \Sigma, \tag{4.1}$$

$$h = \text{the diameter of } K, \tag{4.2}$$

and

$$\rho = \text{the supremum of the diameters of the spheres contained in } K. \tag{4.3}$$

Let $u : \mathbf{R}^n \to \mathbf{R}$ be a function and $\tilde{u} \in \mathcal{P}_k$ a polynomial such that u and $\tilde{u}$ and some corresponding derivatives of u and $\tilde{u}$ are equal at some specified points of Σ. When this problem has a unique solution they proved the error estimate (4.4) below on K. From this we see that the error estimates depend in a crucial way on the geometry of $K = \mathrm{con}(\Sigma)$, in particular, on the quotient h/ρ. If we want to choose Σ as a subset of $F \subset \mathbf{R}^n$ where F preserves Markov's inequality, we see from Property 5 in Section 1.2 that it should be possible to choose Σ so that h/ρ is less than a fixed constant depending only on F.

We start (Section 4.1) by describing Ciarlet-Raviart's result in more detail and by proving an error estimate outside K. Then (Section 4.2) we use this to derive an error estimate for interpolation on any F preserving Markov's inequality and, finally, we give some examples (Section 4.3). We refer to the paper by Ciarlet and Raviart [1] for more details on their results.

4.1. *Interpolation on* $\mathbf{R}^n$

Given a function $u: \mathbf{R}^n \to \mathbf{R}$, its kth (Fréchet) derivative $D^k u(a)$ at $a \in \mathbf{R}^n$, if it exists, is a k-linear mapping from $\mathbf{R}^n \times \cdots \times \mathbf{R}^n$ (k times) to $\mathbf{R}$. The value of $D^k u(a)$ at $(\xi_1, \ldots, \xi_k)$, where $\xi_i \in \mathbf{R}^n$ for $1 \le i \le k$, is denoted by $D^k u(a) \cdot (\xi_1, \ldots, \xi_k)$ and by $D^k u(a) \cdot (\xi)^k$ if $\xi_i = \xi$ for all i. The norm of $D^k u(a)$ is

$$\| D^k u(a) \| = \sup\{ |D^k u(a) \cdot (\xi_1, \ldots, \xi_k)| : |\xi_i| \le 1, 1 \le i \le k \}$$

and satisfies, for $|j| = k$,

$$|D_j u(a)| \le \| D^k u(a) \| \le c(n, k) \max_{|j|=k} |D_j u(a)|.$$

For $k = 0$, $D^k u(a)$ is interpreted as $u(a)$ and $\|D^k u(a)\|$ as $|u(a)|$. We write $Du(a)$ instead of $D^1 u(a)$. Finally, $Du(a) \cdot (\xi)$ is the directional derivative of u at a in the direction ξ, $D^2 u(a) \cdot (\xi_1, \xi_2)$ is the directional derivative of the function $x \to Du(x) \cdot (\xi_2)$ at a in the direction ξ_1, and so on.

Lagrange Interpolation. Here we interpolate the values of a given function at given interpolation points. A set $\Sigma = \{a_1, \ldots, a_N\}$ of distinct points of $\mathbf{R}^n$ (the interpolation points) is called a *k-unisolvent set* if, given any $\alpha_i \in \mathbf{R}$, $1 \le i \le N$, there exists exactly one polynomial $P \in \mathcal{P}_k$ such that $P(a_i) = \alpha_i$, $1 \le i \le N$ (see [1] and Section 4.3 for examples). Before stating the result by Ciarlet-Raviart for Lagrange interpolation we need one more definition. Two k-unisolvent sets $\Sigma = \{a_1, \ldots, a_N\}$ and $\hat{\Sigma} = \{\hat{a}_1, \ldots, \hat{a}_N\}$ are *equivalent* if there exists an invertible linear mapping $T: \mathbf{R}^n \to \mathbf{R}^n$ and $b \in \mathbf{R}^n$ so that $a_i = T\hat{a}_i + b$ for $1 \le i \le N$.

Theorem 4.1 [1, p. 184]. *Let $\Sigma = \{a_i\}_1^N \subset \mathbf{R}^n$ be a k-unisolvent set of points and let K, h, and ρ be defined as in* (4.1)–(4.3). *Let $u \in C^{k+1}(K)$ and let $\tilde{u}$ be the unique polynomial in $\mathcal{P}_k$ interpolating to u at Σ, i.e., $\tilde{u}(a_i) = u(a_i)$, $1 \le i \le N$. Then, for $0 \le m \le k$,*

$$\sup_{x \in K} \| D^m u(x) - D^m \tilde{u}(x) \| \le c \frac{h^{k+1}}{\rho^m} \sup_{x \in K} \| D^{k+1} u(x) \| \tag{4.4}$$

for some constants $c = c(n, k, m, \hat{\Sigma})$ which are the same for all equivalent k-unisolvent sets and which can be computed once and for all for a fixed k-unisolvent set $\hat{\Sigma}$ equivalent to Σ.

As a complement of this result we prove the following error estimate at points y outside K.

Theorem 4.2. *Using the same notation and assumptions as in Theorem* 4.1, *let* $y \in \mathbf{R}^n \backslash K$, *let* $K(y)$ *be the closed convex hull of* Σ *and* y, *let* $d(y, K)$ *denote the distance from* y *to* K, *and assume, furthermore, that* $u \in C^{k+1}(K(y))$. *Put* $M_{k+1}(y) = \sup\{\|D^{k+1}u(x)\|: x \in K(y)\}$. *Then, for* $0 \leq m \leq k$,

$$\|D^m u(y) - D^m \tilde{u}(y)\| \leq c \frac{(h + d(y, K))^{k+1}}{\rho^m} \cdot \max\left\{1, \left(\frac{d(y, K)}{\rho}\right)^{k-m}\right\} \cdot M_{k+1}(y),$$

where, as in Theorem 4.1, $c = c(n, k, m, \hat{\Sigma})$.

Proof. According to Theorem 1 in [1] we get $\|D^m u(y) - D^m \tilde{u}(y)\| \leq c_1 M_{k+1}(y) \cdot (h + d(y, K))^{k+1} \max_{1 \leq i \leq N} \|D^m p_i(y)\|$, $c_1 = c_1(n, k)$, where p_i are the polynomials in $\mathcal{P}_k$ introduced in Theorem 1 in [1]. Hence, it is enough to estimate $\|D^m p_i(y)\|$. We choose $a \in K$ such that $d(y, K) = |y - a|$ and use Taylor's formula

$$D^m p_i(y) = \sum_{\nu=0}^{k-m} \frac{1}{\nu!} D^{m+\nu} p_i(a) \cdot (y - a)^\nu.$$

By the calculation in the proof of Theorem 2 in [1] we conclude that

$$\|D^m p_i(y)\| \leq c_2 \frac{1}{\rho^m} \max\left\{1, \left(\frac{d(y, K)}{\rho}\right)^{k-m}\right\},$$

$c_2 = c_2(n, k, m, \hat{\Sigma})$. By combining these estimates the theorem follows. ■

Example 3. In general, the estimate in Theorem 4.2 cannot be improved. For the case $k = 1$, $m = 0$, Theorem 4.2 says that

$$|u(y) - \tilde{u}(y)| \leq c(h + d(y, K))^2 \max\left\{1, \frac{d(y, K)}{\rho}\right\} M_2(y).$$

Let us take $n = 2$, $x = (x_1, x_2) \in \mathbf{R}^2$, $\Sigma = \{a_1, a_2, a_3\}$ where $a_1 = (0, a)$, $a > 0$, $a_2 = (-1, 0)$, and $a_3 = (1, 0)$, and $\hat{\Sigma} = (\hat{a}_1, \hat{a}_2, \hat{a}_3)$ where $\hat{a}_1 = (0, 1)$, $\hat{a}_2 = a_2$, and $\hat{a}_3 = a_3$. We take $u(x) = 1 - x_1^2$, and find that $\tilde{u}(x) = x_2/a$. Now we study $u(x) - \tilde{u}(x)$ for $x = (0, x_2)$ where x_2 is a fixed negative number. Then $|u(x) - \tilde{u}(x)|$ is of the order of magnitude $|x_2|/a$ when a approaches zero, proving that the factor $\max\{1, d(y, K)/\rho\}$ cannot in general be improved.

Hermite Interpolation. Here we interpolate function values at given points and the values of some directional derivatives of the function at given points. Let the set of interpolation points Σ be

$$\Sigma = \bigcup_{r=0}^{s} \Sigma^r \qquad \text{where} \quad \Sigma^r = \{a_i^r\}_{i=1}^{N_r}$$

for each r is a set of distinct points a_i^r of $\mathbf{R}^n$. To each a_i^r, $1 \leq r \leq s$, we associate a nonempty subset X_i^r of points $(\xi_1, \ldots, \xi_r) \in \mathbf{R}^n \times \cdots \times \mathbf{R}^n$ (r times) where $\xi_i \neq 0$

for all i; X_i^r represents the set of all directions in which we interpolate rth derivates at a_i^r. We put $X=\{X_i^r\}$, $1\le i\le N_r$, $1\le r\le s$. Then we call the set Σ a *k-unisolvent* set with associated sets of directions X if, given any real numbers A_i^0, $1\le i\le N_0$, and any r-linear mappings A_i^r from $\mathbf{R}^n\times\cdots\times\mathbf{R}^n$ (r times) to $\mathbf{R}$, $1\le i\le N_r$, $1\le r\le s$, there exists exactly one $P\in\mathcal{P}_k$ (the interpolating polynomial) such that, for all i and r,

$$P(a_i^0)=A_i^0$$

and

$$D^rP(a_i^r)\cdot(\xi_1,\ldots,\xi_r)=A_i^r\cdot(\xi_1,\ldots,\xi_r) \qquad \text{for all} \quad (\xi_1,\ldots,\xi_r)\in X_i^r.$$

When the interpolation values A_i^0 and A_i^r are given by $u(a_i^0)$ and $D^ru(a_i^r)$, respectively, where u is a function from some subset of $\mathbf{R}^n$ to $\mathbf{R}$, we denote the above interpolating polynomial $P\in\mathcal{P}_k$ by $\tilde{u}$ and call it the polynomial interpolating to u at Σ in the associated directions X.

Now, let Σ be a k-unisolvent set with associated sets of directions X and let $K=\mathrm{con}(\Sigma)$, h, and ρ be defined as in (4.1)-(4.3). Let $u\in C^{k+1}(K)$ and let $\tilde{u}$ be the unique polynomial interpolating to u at Σ in the associated directions X. Then [1, Theorem 4] the estimate (4.4) still holds with the only change that the notion of equivalent k-unisolvent sets must be modified to the corresponding notion where the associated set of directions X are also taken into consideration (see p. 189 of [1]). Also, the analogue of Theorem 4.2 holds with the same proof.

4.2. Interpolation on F

Let $F\subset\mathbf{R}^n$ preserve Markov's inequality. We now consider Lagrange and Hermite interpolation where the set of interpolation points Σ is a subset of F. We pick a ball $B=B(x_0,r)$, $x_0\in F$ and $0<r\le 1$, and points $a_1,\ldots,a_{n+1}\in B(x_0,r)\cap F$ according to Property 5 in Section 1.2. Let Σ be chosen so that

$$\{a_1,\ldots,a_{n+1}\}\subset\Sigma\subset(B(x_0,r)\cap F) \tag{4.5}$$

and so that Σ is a k-unisolvent set for Lagrange interpolation or for Hermite interpolation with an associated set of directions X (see Section 4.3 for examples). Let $K=\mathrm{con}(\Sigma)$, h, and ρ be defined by (4.1)-(4.3). Then $h\le 2r$ and, by Property 5 in Section 1.2, $\rho\ge c^*r$, where $c^*=c^*(F)>0$. Take $u\in C^{k+1}(F\cap B)$ and let $\tilde{u}$ be the unique polynomial in $\mathcal{P}_k$ interpolating to u at Σ (in the associated directions X in case of Hermite interpolation). The function $u\in C^{k+1}(F\cap B)$ can be extended (see Section 2) to a function in $C^{k+1}(\mathbf{R}^n)$ so that (2.3) holds for the function u. Hence, we get the following theorem by Theorems 4.1 and 4.2 in the Lagrange case and by the corresponding results in the Hermite case.

Theorem 4.3. *Assume that $F\subset\mathbf{R}^n$ preserves Markov's inequality and let Σ be a k-unisolvent set satisfying (4.5) where $B=B(x_0,r)$, $x_0\in F$, $0<r\le 1$, and $\{a_1,\ldots,a_{n+1}\}$ are chosen according to Property 5 in Section 1.2. Let K, h, and ρ be defined by (4.1)-(4.3). Assume that $u\in C^{k+1}(F\cap B)$ and let $\tilde{u}\in\mathcal{P}_k$ be the corresponding interpolating polynomial. Then*

$$\sup_{F\cap B}|(D_ju-D_j\tilde{u})(x)|\le cr^{k+1-|j|}\|u;\,C^{k+1}(F\cap B)\|$$

for $0 \le |j| \le k$ and constants $c = c(F, k, |j|, \hat{\Sigma})$ which are the same for all equivalent k-unisolvent sets and can be computed once and for all for a fixed k-unisolvent set $\hat{\Sigma}$ equivalent to Σ. $\|u; C^{k+1}(F \cap B)\|$ is the norm of $u = \{D_j u\}_{|j| \le k+1}$ in $C^{k+1}(F \cap B)$ introduced in Section 2.

4.3. Examples of k-Unisolvent Sets on F

We take $\Sigma = \{a_1, \ldots, a_{n+1}\} \subset F$ where $a_1, \ldots, a_{n+1}$ are, for instance, chosen in accordance with Property 5 in Section 1.2. We first give an example in the Lagrange case.

Example 4. Let $a_1, \ldots, a_{n+1}$ be the vertices of a nondegenerate n-simplex of $\mathbf{R}^n$. If $\Sigma = \{a_1, \ldots, a_{n+1}\}$ then, as is well known, Σ is a 1-unisolvent set for Lagrange interpolation.

Next we give an example in the Hermite case and for simplicity we describe it for $n = 2$. In fact, the construction we use is the one used by Wingren [16] in his work described in the discussion in Section 3.

Example 5. Take $n = 2$, $x = (x_1, x_2)$, any integer $k \ge 1$, and let a_1, a_2, a_3 be the vertices of a nondegenerate 2-simplex of $\mathbf{R}^2$. Let $\Sigma = \bigcup_0^{k-1} \Sigma^r$ where $\Sigma^0 = \{a_1, a_2, a_3\}$ and $\Sigma^r = \{a_1, a_2\}$ for $1 \le r \le k-1$. At a_1 we interpolate corresponding to all the usual partial derivatives up to order $k-1$. This means that we may choose the set X_1^r of points $(\xi_1, \ldots, \xi_r)$ so that each ξ_i may take the values e_1 and e_2 where $\{e_1, e_2\}$ is the standard base in $\mathbf{R}^2$. At a_2 we interpolate corresponding to all the directional derivatives up to order $k-1$ in the direction determined by the line Π through the points a_2 and a_3. This means that X_2^r may be chosen as the set of all points $(\xi_1, \ldots, \xi_r)$ where each ξ_i is equal to a fixed nonzero vector ξ parallel to Π. We prove that Σ is k-unisolvent with an associated set of directions $X = \{X_1^r, X_2^r\}$. We assume that $a_1 = 0$. Then the line Π through a_2 and a_3 does not pass through the origin and we consider the case when it has an equation $x_2 = c_0 + c_1 x_1$, where $c_0 \ne 0$. Now, let Q be the unique polynomial of degree $k-1$ such that $D_j Q(a_1)$, $|j| \le k-1$, have prescribed values. Let a_{i1} be the first coordinate of a_i, $i = 2, 3$. We note that in our case $a_{31} \ne a_{21}$. For given real numbers β_i, $0 \le i \le k-1$, let β be determined so that the polynomial $P_1: \mathbf{R} \to \mathbf{R}$ defined by

$$P_1(x_1) = \sum_{i=0}^{k-1} \frac{\beta_i}{i!}(x_1 - a_{21})^i + \frac{\beta}{k!}(x_1 - a_{21})^k$$

has a prescribed value at $x_1 = a_{31}$. Furthermore, it is easy to check that there exists a unique homogeneous polynomial H in two variables of degree k determined by

$$H(x) = P_1(x_1) - Q(x) \qquad \text{for} \quad x = (x_1, x_2) \in \Pi.$$

Let H be that polynomial and define $P \in \mathcal{P}_k$ by $P = Q + H$. We claim that P interpolates to the given values in the way required. In fact, $D_j P(a_1) = D_j Q(a_1)$,

$|j| \leq k-1$, are prescribed values. On Π we have $P(x) = P_1(x_1)$, i.e., $P(x) = P(x_1, c_0 + c_1x_1) = P_1(x_1)$. If we take $\xi = (1, c_1)$ which is a vector parallel to Π, we find that $P(a_2) = \beta_0$ and $D^r P(a_2) \cdot (\xi)^r = P_1^{(r)}(a_{21}) = \beta_r$, $1 \leq r \leq k-1$, which are given real numbers. This proves the required interpolation at a_2. Finally, by the choice of β above, $P(a_3) = P_1(a_{31})$ is any prescribed value.

It remains to prove that P is unique. If $P^* \in \mathcal{P}_k$ is a polynomial which also interpolates in the way required at a_1, a_2, and a_3 then by the interpolating conditions at a_1, all the coefficients of $P - P^*$ of degree less than k are zero, i.e., $P - P^*$ is homogeneous of degree k. Considered on Π, $p(x_1) := P(x) - P^*(x)$, $x = (x_1, x_2)$, is a polynomial in one variable of degree k which by the interpolation conditions at a_2 and a_3 satisfy $p^{(j)}(a_{21}) = 0$, $0 \leq j \leq k-1$, and $p(a_{31}) = 0$. Since $a_{21} \neq a_{31}$ in the case we consider we conclude that $p \equiv 0$ which gives that $P^* = P$, proving uniqueness.

Acknowledgment. This paper was written while the author was visiting the Institute for Constructive Mathematics, University of South Florida, Tampa, U.S.A.

References

1. P. G. Ciarlet, P. A. Raviart (1972): *General Lagrange and Hermite interpolation in* $\mathbf{R}^n$ *with applications to finite element methods.* Arch. Rational Mech. Anal., **46**:177–199.
2. R. A. DeVore, R. C. Sharpley (1984): *Maximal functions measuring smoothness.* Mem. Amer. Math. Soc., **47**, No. 293.
3. K. J. Falconer (1985): The Geometry of Fractal Sets. Cambridge: Cambridge University Press.
4. G. Glaeser (1958): *Étude de quelques algèbres Tayloriennes.* J. Analyse Math., **6**:1–125.
5. A. Jonsson (1988): *Markov's inequality and local polynomial approximation.* In: Proceedings of the Seminar on Function Spaces and Applications, Lund, 1986 (M. Cwikel, J. Peetre, Y. Sagher, H. Wallin, eds.). Lecture Notes in Mathematics, vol. 1302. Berlin: Springer-Verlag, pp. 303–316.
6. A. Jonsson, H. Wallin (1979): *The trace to closed sets of functions in* $\mathbf{R}^n$ *with second difference of order* $O(h)$. J. Approx. Theory, **26**:159–184.
7. A. Jonsson, H. Wallin (1984): Function Spaces on Subsets of $\mathbf{R}^n$. Mathematical Reports 2, Part 1. London: Harwood.
8. A. Jonsson, P. Sjögren, H. Wallin (1984): *Hardy and Lipschitz spaces on subsets of* $\mathbf{R}^n$. Studia Math., **80**:141–166.
9. B. B. Mandelbrot (1982): The Fractal Geometry of Nature. San Francisco: Freeman.
10. E. M. Stein (1970): Singular Integrals and Differentiability Properties of Functions. Princeton: Princeton University Press.
11. J. C. Tougeron (1972): Ideaux de fonctions differentiables. Berlin: Springer-Verlag.
12. A. L. Volberg, S. V. Konyagin (1984): *There is a homogeneous measure on any compact subset in* $\mathbf{R}^n$. Dokl. Akad. Nauk SSSR, **278**, No. 4. English transl.: Soviet Math. Dokl., **30**:453–456.
13. H. Wallin (1983): *Markov's inequality on subsets of* $\mathbf{R}^n$. Canad. Math. Soc. Conf. Proc., **3**:377–388.
14. H. Wallin (1988): *New and old function spaces.* In: Proceedings of the Seminar on Function Spaces and Applications, Lund, 1986 (M. Cwikel, J. Peetre, Y. Sagher, H. Wallin, eds.). Lecture Notes in Mathematics, vol. 1302. Berlin: Springer-Verlag, pp. 99–114.
15. H. Whitney (1934): *Analytic extensions of differentiable functions defined in closed sets.* Trans. Amer. Math. Soc., **36**:63–89.

16. P. WINGREN (1988): *Lipschitz spaces and interpolating polynomials on subsets of Euclidean space.* In: Proceedings of the Seminar on Function Spaces and Applications, Lund 1986 (M. Cwikel, J. Peetre, Y. Sagher, H. Wallin, eds.). Lecture Notes in Mathematics, vol. 1302. Berlin: Springer-Verlag, pp. 424–435.
17. P. WINGREN, L. ÖDLUND (1985): Local Polynomial Approximation of Families of Functions in $C^k(F)$. Report No. 4, Department of Math., University of Umeå.

H. Wallin
Department of Mathematics
University of Umeå
S-90187 Umeå
Sweden

Constr. Approx. (1989) 5: 151–170

CONSTRUCTIVE APPROXIMATION

Newton's Method for Fractal Approximation

Wm. Douglas Withers

Abstract. The problem of fitting a given function in the L^q norm with a function generated by an iterated function system can be rapidly solved by applying Newton's method on the parameter space of the iterated function system. The key to this is a method for calculating the derivatives of a potential function with respect to the parameters.

1. Introduction

Many of the objects found in nature have the property that each part is similar to the whole; for example, a branch of a tree may resemble the entire tree, or a small section of a coastline seen at a large scale looks exactly like a large section of coastline seen at a small scale. Such objects can be modeled by *fractals*, one of whose properties is exactly this similarity of parts to the whole. A survey of the essential properties of fractals and of many of the occurrences of fractals in nature is given in Mandelbrot [10]. Some objects in particular can be modeled by *fractal functions*, that is, functions whose graphs are fractal sets. For example a range of mountains or any type of terrain can be considered as a function from $\mathbf{R}^2$ to $\mathbf{R}$, where the independent variables are latitude and longitude and the dependent variable is altitude. Conversely, a river may be modeled as a function from $\mathbf{R}$ to $\mathbf{R}^2$, where the independent variable is altitude and the dependent variables are latitude and longitude.

The *iterated function system* is a convenient tool for generating fractals with specified self-similarity properties. An iterated function system consists of several mappings $f^1, \ldots, f^N$ from a set X to itself with associated probabilities $p^1, \ldots, p^N$. Successive states of the system are generated by the rule that if x^t is the state of the system at time t, then the state x^{t+1} at time $t+1$ is given by $x^{t+1} = f^n(x^t)$ with probability p^n. For a large class of these systems, the successive states approach a compact set, the *attractor* for the system, which is completely independent of the choice of initial state x^0. The mappings f^n correspond to the similarities which we want the attractor to have. For example, if we want the attractor to represent a tree and we want each branch of the tree to resemble the

Date received: December 5, 1986. Date revised: January 20, 1988. Communicated by Michael F. Barnsley.
AMS classification: 41A30, 60J05, 85F11.
Key words and phrases: Fractal interpolation, Iterated function system, Newton's method.

tree as a whole, perhaps flattened or otherwise proportioned differently, we would define one mapping f^n to be that which takes the entire tree onto one branch. An introduction to the use of iterated function systems to generate approximations to fractal objects is given in Barnsley [1] and Barnsley and Demko [2].

In some situations we may wish to find the best approximation possible in the L^q norm to an object by a fractal function whose graph is the attractor for an iterated function system dependent on several parameters. For example, in modeling terrain for a flight simulator, it is essential that the large-scale features of the terrain be correct, which can be achieved by making the approximation close in the L^q norm. At the same time it is desirable to make the small-scale details look right. The approximation given by an iterated function system is especially appropriate for this because the fractal dimension of the attractor can be adjusted; see, for example, the formulas of Hardin and Massopoust [8] for the capacity of the attractor for iterated function systems in two dimensions, generalized in Barnsley, Elton, Hardin, and Massopoust [6]. In this article we consider the problem of finding this best approximation.

Our approach is to define a "potential" function of the parameters for the system, based on the L^q norm of the difference between the target function and the function generated by the iterated function system. We then seek a minimum for the potential function, by applying Newton's method to the gradient. A key component of the method is thus a technique for calculating the derivative of the potential with respect to the parameters of the system. We have considered the questions of differentiability and calculation of derivatives with respect to parameters in iterated function systems previously in [12], [13], and [14]. In practice our method is an efficient way to fit a target function with a function generated by an iterated function system.

The structure of this paper is as follows. In Section 2 we describe the background of our problem and state our theorems. The proofs of these theorems are then developed in the following three sections. In Section 3 we describe Newton's method as applied to the gradient of the potential. In Section 4 we discuss the symbolic analogue of the iterated function system. In Section 5 we discuss the problem of taking derivatives of the potential with respect to the parameters of the system. In Section 6 we present an example of our method as applied to a particular system consisting of several affine maps in the plane.

2. Background and Statement of Theorem

Let $\mathbf{u}$ be the function which we want to approximate. We let D be the domain of $\mathbf{u}$ and R a set containing its range. We assume D and R are compact subsets of $\mathbf{R}^L$ and $\mathbf{R}^J$, respectively, which are the closures of their interiors.

We want to approximate $\mathbf{u}$ in the L^q norm with a function $\mathbf{v}$ generated by an iterated function system $\mathbf{f}^1, \ldots, \mathbf{f}^N: \mathbf{R}^{L+J} \to \mathbf{R}^{L+J}$, with associated probabilities $p^1, \ldots, p^N: \mathbf{R}^L \to (0, 1)$, so that if the position at time t is given by $\mathbf{z}^t = (\mathbf{x}^t, \mathbf{y}^t)$, $\mathbf{x} \in \mathbf{R}^L$, $\mathbf{y} \in \mathbf{R}^J$, then the position $\mathbf{z}^{t+1}$ at time $t+1$ is given by

$$\mathbf{z}^{t+1} = \mathbf{f}^n(\mathbf{z}^t)$$

with probability $p^n(\mathbf{x}')$. The attractor Γ for the system will be the graph $\{(\mathbf{x}, \mathbf{v}(\mathbf{x})): \mathbf{x} \in D\}$ of the function $\mathbf{v}$.

In order that the attractor be the graph of a function except on a set of measure zero, we require each $\mathbf{f}^n$ to be a *generalized shear*; that is, there exist $\mathbf{g}: \mathbf{R}^L \to \mathbf{R}^L$ and $\mathbf{h}: \mathbf{R}^{L+J} \to \mathbf{R}^J$ such that

$$\mathbf{f}^n(\mathbf{x}, \mathbf{y}) = (\mathbf{g}^n(\mathbf{x}), \mathbf{h}^n(\mathbf{x}, \mathbf{y})).$$

Moreover, we require that each $\mathbf{g}^n$ be injective on D and that

$$D = \bigcup_{n=1}^{N} \mathbf{g}^n(D),$$

the sets $\mathbf{g}^n(D)$ overlapping only on a set of Lebesgue measure zero.

We assume that the mappings $\mathbf{h}^n$ depend on parameters $w_1, \ldots, w_K$, although we often suppress dependence on $w_1, \ldots, w_K$ in our notation. To simplify our problem, we assume that the functions $\mathbf{g}^n$ and the probabilities p^n are independent of $\mathbf{w}$. The function $\mathbf{v}$ then depends on $\mathbf{w} = (w_1, \ldots, w_K)$ as well as $\mathbf{x} = (x_1, \ldots, x_L)$. Our problem is thus to find the point $\mathbf{w}$ in parameter space which minimizes

$$V(\mathbf{w}) = \int_D \|\mathbf{v}(\mathbf{x}, \mathbf{w}) - \mathbf{u}(\mathbf{x})\|^q \, d\mu(\mathbf{x}), \tag{2.1}$$

where $\|\cdot\|$ is the L^q norm on $\mathbf{R}^J$ and μ is normalized Lebesgue measure on D. We assume that such a minimum exists and we denote by $\mathbf{w}^*$ the point at which V takes its minimum.

If the mappings $\mathbf{f}^n$ are contracting and the probabilities p^n are Lipschitz-continuous, then the iterated function system $\mathbf{f}^1, \ldots, \mathbf{f}^N$ generates a unique probability measure ν supported on Γ which is invariant under the system:

$$\nu \mathbf{f}^n(E) = \int_E p^n(\mathbf{z}) \, d\nu(\mathbf{z}).$$

This is proven under weaker hypotheses in Barnsley and Demko [2] and elsewhere [3]-[5]. The hypotheses of our theorems will ensure the existence of a unique invariant probability measure. So that we can calculate integrals similar to that in (2.1) as trajectory averages, we want the projection onto D of the invariant measure ν generated by the iterated function system to equal μ; in other words, for a measurable set $E \subset D$, we have $\nu(E \times R) = \mu E$. We therefore require

$$\sum_{n=1}^{N} |\det \mathbf{J}(\mathbf{g}^n)| = 1,$$

where $\det \mathbf{J}$ is the Jacobian determinant, and

$$p^n = |\det \mathbf{J}(\mathbf{g}^n)|.$$

The nature of our algorithm is given by the following two theorems:

Theorem 2.1. *Suppose V is continuously twice-differentiable and its matrix of second derivatives is nonsingular at $\mathbf{w}^*$, hence in a neighborhood O of $\mathbf{w}^*$. For*

$\mathbf{w}\in O$, let $\alpha_k(\mathbf{w})=\partial V/\partial w_k(\mathbf{w})$ and $\beta_{kl}(\mathbf{w})=\partial^2 V/\partial w_k\,\partial w_l(\mathbf{w})$. Let $[\gamma_{kl}(\mathbf{w})]$ be the inverse matrix of $[\beta_{kl}(\mathbf{w})]$. Define $\mathbf{r}$: $O\to\mathbf{R}^K$ by

$$r_k(\mathbf{w})=w_k-\sum_{l=1}^{K}\gamma_{kl}(\mathbf{w})\alpha_l(\mathbf{w}),$$

where r_k is the kth component of $\mathbf{r}$. Then $\mathbf{w}^*$ is a superstable fixed point for the mapping $\mathbf{r}$; thus for all initial points $\mathbf{w}$ in some neighborhood of $\mathbf{w}^*$, the iterates of $\mathbf{w}$ under $\mathbf{r}$ converge rapidly to $\mathbf{w}^*$.

Theorem 2.2. *Suppose $\mathbf{f}^1,\ldots,\mathbf{f}^N$ are contractions and have Lipschitz-continuous second derivatives in $D\times R$. Let $\mathbf{z}^0=(\mathbf{x}^0,\mathbf{y}^0)$, $a^0_{j;k}$, and $b^0_{j;kl}$, $j=1,\ldots,J$; $k,l=1,\ldots,K$, be chosen arbitrarily and let $\mathbf{z}^t=(\mathbf{x}^t,\mathbf{y}^t)$, $a^t_{j;k}$, and $b^t_{j;kl}$ be defined iteratively by the following process. Choose n with probability $p^n(\mathbf{x}^t)$ and let*

$$\mathbf{z}^{t+1}=\mathbf{f}^n(\mathbf{z}^t),$$

$$a^{t+1}_{j;k}=\sum_{i=1}^{J}\frac{\partial h^n_j}{\partial y_i}(\mathbf{z}^t)a^t_{i;k}+\frac{\partial h^n_j}{\partial w_k}(\mathbf{z}^t), \tag{2.2}$$

$$\begin{aligned} b^{t+1}_{j;kl}={}&\sum_{i=1}^{J}\sum_{m=1}^{J}\frac{\partial^2 h^n_j}{\partial y_i\,\partial y_m}(\mathbf{z}^t)a^t_{i;k}a^t_{m;l}+\sum_{i=1}^{J}\frac{\partial^2 h^n_j}{\partial y_i\,\partial w_l}(\mathbf{z}^t)a^t_{i;k}\\ &+\sum_{i=1}^{J}\frac{\partial^2 h^n_j}{\partial y_i\,\partial w_k}(\mathbf{z}^t)a^t_{i;l}+\frac{\partial^2 h^n_j}{\partial w_k\,\partial w_l}(\mathbf{z}^t)+\sum_{i=1}^{J}\frac{\partial h^n_j}{\partial y_i}(\mathbf{z}^t)b^t_{i;kl}, \end{aligned} \tag{2.3}$$

where h^n_j is the j*th component of $\mathbf{h}^n$. Then with probability one,*

$$\frac{\partial V}{\partial w_k}=\lim_{T\to\infty}\frac{1}{T}\sum_{t=1}^{T}\sum_{j=1}^{J}q\xi|y^t_j-u_j(\mathbf{x}^t)|^{q-1}a^t_{j;k} \tag{2.4}$$

and

$$\frac{\partial^2 V}{\partial w_k\,\partial w_l}=\lim_{T\to\infty}\frac{1}{T}\sum_{t=1}^{T}[q(q-1)|y^t_j-u_j(\mathbf{x}^t)|^{q-2}a^t_{j;k}a^t_{j;l}+q\xi|y^t_j-u_j(\mathbf{x}^t)|^{q-1}b^t_{j;kl}], \tag{2.5}$$

where y^t_j and u_j are the j*th components of $\mathbf{y}^t$ and $\mathbf{u}$, respectively; and $\xi=\operatorname{sgn}(y^t_j-u_j(\mathbf{x}^t))$.*

The proof of Theorem 2.1 is given in Section 3 and that of Theorem 2.2 is given in Section 5.

3. Newton's Method

One way to find a value $\mathbf{w}^*$ of $\mathbf{w}$ which minimizes $V(\mathbf{w})$ is to apply Newton's method to the gradient of V with respect to $\mathbf{w}$. This is an iterative method; in other words, given an approximation $\mathbf{w}$ to $\mathbf{w}^*$, the method will generate a better approximation $\mathbf{w}'$ to $\mathbf{w}^*$. Iteration of this process should give a sequence of values which converges rapidly to $\mathbf{w}^*$, since, as we show in this section, $\mathbf{w}^*$ is a superstable fixed point for Newton's method.

Given an approximation $\mathbf{w}=(w_1,\ldots,w_K)$, there are two steps in finding the next approximation $\mathbf{w}'=(w_1',\ldots,w_K')$:

(i) Take first and second derivatives of V with respect to $w_1,\ldots,w_K$.
(ii) Construct the quadratic approximation to V near $(w_1,\ldots,w_K)$ and let $(w_1',\ldots,w_K')$ be the point at which the gradient of the quadratic approximation is zero.

In this section we discuss part (ii) of this process. In Section 3 we deal with the symbolic analogue of the iterated function system and in Section 4 we discuss the problem of calculating derivatives with respect to $(w_1,\ldots,w_K)$.

Let $\alpha_k(\mathbf{w})=\partial V/\partial w_k(\mathbf{w})$ and $\beta_{kl}(\mathbf{w})=\partial^2 V/\partial w_k\,\partial w_l(\mathbf{w})$, $k,l=1,\ldots,K$. Then the quadratic approximation Q to V near $\mathbf{w}$ is given by

$$Q(\mathbf{w}+\Delta\mathbf{w})=V(\mathbf{w})+\sum_{k=1}^{K}\alpha_k(\mathbf{w})\Delta w_k+\tfrac{1}{2}\sum_{k=1}^{K}\sum_{l=1}^{K}\Delta w_k\beta_{kl}(\mathbf{w})\Delta w_l.$$

We thus have

$$\frac{\partial Q}{\partial w_k}(\mathbf{w})=\alpha_k(\mathbf{w})+\sum_{l=1}^{K}\beta_{kl}(\mathbf{w})\Delta w_l.$$

To find the next value of $\mathbf{w}$ we set all these to zero and solve the resulting system of linear equations for $\Delta\mathbf{w}=(\Delta w_1,\ldots,\Delta w_K)$. We then have $\mathbf{w}'=\mathbf{w}+\Delta\mathbf{w}$.

Assuming that the matrix $[\beta_{kl}(\mathbf{w})]$ is nonsingular, let its inverse matrix be $[\gamma_{kl}(\mathbf{w})]$:

$$\sum_{k=1}^{K}\gamma_{jk}(\mathbf{w})\beta_{kl}(\mathbf{w})=\delta_{jl},$$

where δ_{jl} is the Kronecker delta ($\delta_{jl}=1$ if $j=l$, 0 otherwise). We then have

$$w_k'=w_k-\sum_{l=1}^{K}\gamma_{kl}(\mathbf{w})\alpha_l(\mathbf{w})=r_k(\mathbf{w}),$$

where $r_k(\mathbf{w})$ is the kth component of $\mathbf{r}(\mathbf{w})$.

Proof of Theorem 2.1. Since $\alpha_k(\mathbf{w}^*)=0$, $\mathbf{w}^*$ is a fixed point for $\mathbf{r}$. Let us calculate the derivative matrix of the mapping $\mathbf{r}(\mathbf{w})$:

$$\begin{aligned}\frac{\partial r_k}{\partial w_j}(\mathbf{w})&=\delta_{jk}-\sum_{l=1}^{K}\frac{\partial\gamma_{kl}}{\partial w_j}(\mathbf{w})\alpha_l(\mathbf{w})-\sum_{l=1}^{K}\gamma_{kl}(\mathbf{w})\frac{\partial\alpha_l}{\partial w_j}(\mathbf{w}).\\&=\delta_{jk}-\sum_{l=1}^{K}\frac{\partial\gamma_{kl}}{\partial w_j}(\mathbf{w})\alpha_l(\mathbf{w})-\sum_{l=1}^{K}\gamma_{kl}(\mathbf{w})\beta_{lj}(\mathbf{w})\\&=-\sum_{l=1}^{K}\frac{\partial\gamma_{kl}}{\partial w_j}(\mathbf{w})\alpha_l(\mathbf{w}).\end{aligned}$$

Since $\alpha_l(\mathbf{w}^*)=0$, we have

$$\frac{\partial r_k}{\partial w_j}(\mathbf{w}^*)=0.$$

The point $\mathbf{w}^*$ is thus a superstable fixed point for $\mathbf{r}$. It follows that all values of $\mathbf{w}$ in some neighborhood of $\mathbf{w}^*$ will rapidly approach $\mathbf{w}^*$ under iteration of the mapping $\mathbf{r}$. ■

4. The Symbolic Iterated Function System

Symbolic dynamics is a common technique for analysis of dynamical systems and iterated function systems are especially open to this approach. The basic idea is to draw an analogy between the iterated function system on $D \times R$ and a system of shifts on sequences of symbols. Let Σ be the set of infinite sequences from $\{1, \ldots, N\}$. Let s^n be the nth right shift on Σ:

$$s^n(n_1, n_2, \ldots) = (n, n_1, n_2, \ldots).$$

We can make Σ into a metric space as follows. Let $\mathbf{n} = (n_1, n_2, \ldots)$ and $\mathbf{n}' = (n_1', n_2', \ldots)$ be elements of Σ. Then we define

$$d(\mathbf{n}, \mathbf{n}') = \sum_{k=1}^{\infty} (1 - \delta_{n_k n_k'}) 2^{-k}.$$

With this metric, each s^n is contracting with constant $\frac{1}{2}$ and Σ is a compact space with a Cantor set topology.

Lemma 4.1. *Let C be a compact subset of $\mathbf{R}^M$ and let $\mathbf{F}^1, \ldots, \mathbf{F}^N$ be contractions from C to C; that is, there exists $\lambda \in (0, 1)$ such that*

$$\|\mathbf{F}^n(\mathbf{z}) - \mathbf{F}^n(\mathbf{z}')\| \le \lambda \|\mathbf{z} - \mathbf{z}'\|.$$

For $k = 1, 2, \ldots$ define $\mathbf{\Phi}_k : C \times \Sigma \to C$ by

$$\mathbf{\Phi}_k(\mathbf{z}, \mathbf{n}) = \mathbf{F}^{n_1} \circ \mathbf{F}^{n_2} \circ \cdots \circ \mathbf{F}^{n_k}(\mathbf{z}), \tag{4.1}$$

where $\mathbf{n} = (n_1, n_2, \ldots)$. Then $\mathbf{\Phi}_k(\mathbf{z}, \mathbf{n})$ converges uniformly to a value independent of $\mathbf{z}$ and continuous in $\mathbf{n}$ as $k \to \infty$. If $\mathbf{F}^1, \ldots, \mathbf{F}^N$ also depend on a parameter $\mathbf{w}$ in some subset of $\mathbf{R}^K$ and λ is independent of $\mathbf{w}$, then this convergence is uniform in $\mathbf{w}$.

Proof. It is straightforward to show by induction that, for fixed $\mathbf{n}$, $\mathbf{\Phi}_k(\mathbf{z}, \mathbf{n})$ is contracting in $\mathbf{z}$ with constant λ^k. We then have, for $j > k$,

$$\begin{aligned} \|\mathbf{\Phi}_j(\mathbf{z}, \mathbf{n}) - \mathbf{\Phi}_k(\mathbf{z}, \mathbf{n})\| &= \|\mathbf{\Phi}_k(\mathbf{F}^{n_{k+1}} \circ \cdots \circ \mathbf{F}^{n_j}(\mathbf{z}), \mathbf{n}) - \mathbf{\Phi}_k(\mathbf{z}, \mathbf{n})\| \\ &\le \lambda^k \|\mathbf{F}^{n_{k+1}} \circ \cdots \circ \mathbf{F}^{n_j}(\mathbf{z}) - \mathbf{z}\| \\ &\le \lambda^k \operatorname{diam} C; \end{aligned} \tag{4.2}$$

thus the sequence $\mathbf{\Phi}_k(\mathbf{z}, \mathbf{n})$ is uniformly Cauchy and therefore uniformly convergent. Let

$$\mathbf{\Phi}(\mathbf{z}, \mathbf{n}) = \lim_{k \to \infty} \mathbf{\Phi}_k(\mathbf{z}, \mathbf{n}).$$

Letting $j \to \infty$ in (4.2) we find that the convergence is uniform in $\mathbf{z}$ and $\mathbf{n}$:

$$\|\mathbf{\Phi}(\mathbf{z}, \mathbf{n}) - \mathbf{\Phi}_k(\mathbf{z}, \mathbf{n})\| \le \lambda^k \operatorname{diam} C. \tag{4.3}$$

Moreover, since

$$\sup_{\mathbf{z},\mathbf{z}'\in C} \|\mathbf{\Phi}_k(\mathbf{z},\mathbf{n})-\mathbf{\Phi}_k(\mathbf{z}',\mathbf{n})\| \le \lambda^k \operatorname{diam} C \to 0$$

as $k\to\infty$, $\mathbf{\Phi}(\mathbf{z},\mathbf{n})=\mathbf{\Phi}(\mathbf{n})$ is independent of $\mathbf{z}$.

If $\mathbf{n},\mathbf{n}'\in\Sigma$ and $d(\mathbf{n},\mathbf{n}')<2^{-k}$, then $n_j=n_j'$ for $j=1,\dots,k$. Thus, for $j>k$,

$$\begin{aligned}\|\mathbf{\Phi}_j(\mathbf{z},\mathbf{n})-\mathbf{\Phi}_j(\mathbf{z},\mathbf{n}')\| &= \|\mathbf{F}^{n_1}\circ\cdots\circ\mathbf{F}^{n_k}(\mathbf{F}^{n_{k+1}}\circ\cdots\circ\mathbf{F}^{n_j}(\mathbf{z}))\\ &\quad -\mathbf{F}^{n_1}\circ\cdots\circ\mathbf{F}^{n_k}(\mathbf{F}^{n'_{k+1}}\circ\cdots\circ\mathbf{F}^{n'_j}(\mathbf{z}))\|\\ &\le \lambda^k \operatorname{diam} C.\end{aligned}$$

Letting $j\to\infty$, we obtain

$$\|\mathbf{\Phi}(\mathbf{n})-\mathbf{\Phi}(\mathbf{n}')\|\le\lambda^k \operatorname{diam} C,$$

which shows that $\mathbf{\Phi}$ is continuous in $\mathbf{n}$.

In the case where $\mathbf{F}^1,\dots,\mathbf{F}^N$ depend on $\mathbf{w}$ as well as $\mathbf{z}$, but λ is independent of $\mathbf{w}$, (4.3) still holds; thus the convergence is uniform in $\mathbf{w}$ as well as $\mathbf{z}$. ■

Letting $\mathbf{F}^n=\mathbf{f}^n$ in this lemma, we can conclude that corresponding to each $\mathbf{n}\in\Sigma$ is a point $\boldsymbol{\varphi}(\mathbf{n})=\boldsymbol{\varphi}(\mathbf{n},\mathbf{w})$ in Γ given by

$$\begin{aligned}\boldsymbol{\varphi}(n_0,n_1,\dots)&=\mathbf{f}^{n_0}\circ\mathbf{f}^{n_1}\circ\cdots\\ &=\lim_{k\to\infty}\mathbf{f}^{n_0}\circ\mathbf{f}^{n_1}\circ\cdots\circ\mathbf{f}^{n_k}(D\times R).\end{aligned}$$

We then have

$$\mathbf{f}^n(\boldsymbol{\varphi}(\mathbf{n}))=\boldsymbol{\varphi}(s^n(\mathbf{n})).$$

In other words, $\boldsymbol{\varphi}$ satisfies the functional equation

$$\boldsymbol{\varphi}\circ s^n=\mathbf{f}^n\circ\boldsymbol{\varphi}.$$

Note that $\boldsymbol{\varphi}(\Sigma)$ is the graph of $\mathbf{v}$.

For each $\mathbf{n}\in\Sigma$ we can also define a point $\boldsymbol{\psi}(\mathbf{n})$ in D given by

$$\begin{aligned}\boldsymbol{\psi}(n_0,n_1,\dots)&=\mathbf{g}^{n_0}\circ\mathbf{g}^{n_1}\circ\cdots\\ &=\lim_{k\to\infty}\mathbf{g}^{n_0}\circ\mathbf{g}^{n_1}\circ\cdots\circ\mathbf{g}^{n_k}(D).\end{aligned}$$

We thus have $\boldsymbol{\psi}(\Sigma)=D$. We also define

$$\begin{aligned}\boldsymbol{\chi}(n_0,n_1,\dots)=\boldsymbol{\chi}(n_0,n_1,\dots,\mathbf{w})&=\mathbf{h}^{n_0}\circ\mathbf{f}^{n_1}\circ\cdots\\ &=\lim_{k\to\infty}\mathbf{h}^{n_0}\circ\mathbf{f}^{n_1}\circ\cdots\circ\mathbf{f}^{n_k}(D\times R),\end{aligned}$$

so that $\boldsymbol{\varphi}(\mathbf{n})=(\boldsymbol{\psi}(\mathbf{n}),\boldsymbol{\chi}(\mathbf{n}))$ and $\mathbf{v}(\boldsymbol{\psi}(\mathbf{n}))=\boldsymbol{\chi}(\mathbf{n})$.

We then have the following functional equations:

$$\boldsymbol{\psi}\circ s^n=\mathbf{g}^n\circ\boldsymbol{\psi}$$

and

$$\boldsymbol{\chi}\circ s^n=\mathbf{h}^n\circ\boldsymbol{\varphi}. \tag{4.4}$$

We can now make an iterated function system on Σ consisting of the mappings $s^1, \ldots, s^N$. If the state of the system at time t is a sequence $\mathbf{n}^t$, then the state at time $t+1$ is the sequence $\mathbf{n}^{t+1} = s^n(\mathbf{n}^t)$ with probability $p^n(\boldsymbol{\psi}(\mathbf{n}^t))$. It can be shown that there is a unique probability measure σ on Σ which is invariant under this system:

$$\sigma(s^n(E)) = \int_E p^n(\boldsymbol{\psi}(\mathbf{n}))\, d\sigma(\mathbf{n}).$$

The measure σ corresponds to the measure ν on $D\times R$ which is invariant under (1.2), ν being given by

$$\nu E = \sigma(\boldsymbol{\varphi}^{-1}(E)).$$

The reason for the condition that the functions $\mathbf{g}^n$ and the probabilities p^n not depend on $\mathbf{w}$ or $\mathbf{y}$ is to ensure that σ is independent of $\mathbf{w}$. This greatly simplifies our problem.

Lemma 4.2. *Assume* $\mathbf{f}^1, \ldots, \mathbf{f}^N$ *are contractions in* $\mathbf{z}$; *there exists* $\lambda \in (0, 1)$ *such that*

$$\|\mathbf{f}^n(\mathbf{z}, \mathbf{w}) - \mathbf{f}^n(\mathbf{z}', \mathbf{w})\| \le \lambda \|\mathbf{z}-\mathbf{z}'\|$$

for $\mathbf{z}, \mathbf{z}' \in D\times R$, λ *being independent of* $\mathbf{w}$ *for* $\mathbf{w}$ *in some compact subset* S *of* $\mathbf{R}^k$. *Assume further that* $\mathbf{f}^1, \ldots, \mathbf{f}^N$ *have Lipschitz-continuous second derivatives with respect to* $\mathbf{z}$ *and* $\mathbf{w}$ *for* $\mathbf{z}\in D\times R$ *and* $\mathbf{w}\in S$. *Then, for fixed* $\mathbf{n}\in\Sigma$, $\boldsymbol{\varphi} = \boldsymbol{\varphi}(\mathbf{n}, \mathbf{w})$ *is continuously twice-differentiable in* $\mathbf{w}$ *for* $\mathbf{w}$ *in the interior of* S *and its second derivatives are continuous functions of* $\mathbf{n}$.

Proof. First assume $\mathbf{f}^1, \ldots, \mathbf{f}^N$ to have Lipschitz-continuous first derivatives and let w be one of $w_1, \ldots, w_K$; we will show that $\partial\boldsymbol{\varphi}/\partial w$ exists. Let $\boldsymbol{\varphi}_k$ be the function $\boldsymbol{\Phi}_k$ defined by equation (4.1) with $\mathbf{F}^n = \mathbf{f}^n$. Since $\boldsymbol{\varphi} = \lim \boldsymbol{\varphi}_k$, and $\boldsymbol{\varphi}_k$ is continuously differentiable in $\mathbf{w}$, it is sufficient to show that $\partial\boldsymbol{\varphi}_k/\partial w$ converges uniformly in $\mathbf{w}$. To this end we define an iterated function system of larger dimension as follows.

Let Λ be a Lipschitz constant for $\mathbf{Df}^n$ and $\partial\mathbf{f}^n$ and $\partial\mathbf{f}^n/\partial w$ for all fixed $\mathbf{w}\in S$, where $\mathbf{Df}^n(\mathbf{z})$ is the derivative matrix with respect to $\mathbf{z}$ for $\mathbf{f}^n$ at $\mathbf{z}$:

$$\left\|\frac{\partial \mathbf{f}^n}{\partial w}(\mathbf{z}) - \frac{\partial \mathbf{f}^n}{\partial w}(\mathbf{z}')\right\| < \Lambda\|\mathbf{z}-\mathbf{z}'\|,$$

$$\|\mathbf{Df}^n(\mathbf{z}) - \mathbf{Df}^n(\mathbf{z}')\| < \Lambda\|\mathbf{z}-\mathbf{z}'\|.$$

Let M be a bound on $\partial\mathbf{f}^1/\partial w, \ldots, \partial\mathbf{f}^N/\partial w$ for $\mathbf{w}\in S$, $\mathbf{z}\in D\times R$. By the contracting property of $\mathbf{f}^n$, $\|\mathbf{Df}^n(\mathbf{z})\| < \lambda$. Let

$$\eta = \frac{(1-\lambda)^2}{2\Lambda(1-\lambda+M)}.$$

Let B be the closed ball centered at the origin of $\mathbf{R}^{J+L}$ with radius $\eta M/(1-\lambda)$. Define $\mathbf{F}^1, \ldots, \mathbf{F}^N : D\times R\times B \to D\times R\times B$ by

$$\mathbf{F}^n(\mathbf{z}, \mathbf{a}) = \left(\mathbf{f}^n(\mathbf{z}), \mathbf{Df}^n(\mathbf{z})\mathbf{a} + \eta\frac{\partial \mathbf{f}^n}{\partial w}(\mathbf{z})\right) \tag{4.5}$$

for $\mathbf{z}\in D\times R$, $\mathbf{a}\in B$. It is easily verified that $\mathbf{F}^n$ does indeed map $D\times R\times B$ to itself.

Then each $\mathbf{F}^n$ is a contraction:

$$\|\mathbf{F}^n(\mathbf{z},\mathbf{a})-\mathbf{F}^n(\mathbf{z}',\mathbf{a}')\| = \left\|\left(\mathbf{f}^n(\mathbf{z})-\mathbf{f}^n(\mathbf{z}'), D\mathbf{f}^n(\mathbf{z})\mathbf{a}-D\mathbf{f}^n(\mathbf{z}')\mathbf{a}'+\eta\frac{\partial\mathbf{f}^n}{\partial w}(\mathbf{z})-\eta\frac{\partial\mathbf{f}^n}{\partial w}(\mathbf{z}')\right)\right\|$$

$$\le \|(\mathbf{f}^n(\mathbf{z})-\mathbf{f}^n(\mathbf{z}'), D\mathbf{f}^n(\mathbf{z})\mathbf{a}-D\mathbf{f}^n(\mathbf{z})\mathbf{a}')\|$$

$$+\|(0, D\mathbf{f}^n(\mathbf{z})\mathbf{a}'-D\mathbf{f}^n(\mathbf{z}')\mathbf{a}')\|$$

$$+\left\|\left(0, \eta\frac{\partial\mathbf{f}^n}{\partial w}(\mathbf{z})-\eta\frac{\partial\mathbf{f}^n}{\partial w}(\mathbf{z}')\right)\right\|$$

$$\le \lambda\|(\mathbf{z},\mathbf{a})-(\mathbf{z}',\mathbf{a}')\|+\Lambda\|\mathbf{z}-\mathbf{z}'\|\,\|\mathbf{a}'\|+\Lambda\eta\|\mathbf{z}-\mathbf{z}'\|$$

$$\le\left(\lambda+\Lambda\frac{\eta M}{1-\lambda}+\Lambda\eta\right)\|(\mathbf{z},\mathbf{a})-(\mathbf{z}',\mathbf{a}')\|$$

$$\le\frac{\lambda+1}{2}\|(\mathbf{z},\mathbf{a})-(\mathbf{z}',\mathbf{a}')\|.$$

We can therefore apply Lemma 4.1 to this system; construct $\mathbf{\Phi}_k$ as in equation (4.1) and $\mathbf{\Phi}_k(\mathbf{z},\mathbf{n})$ converges uniformly to a value independent of $\mathbf{z}$ and continuous in $\mathbf{n}$ as $k\to\infty$.

Using simple calculus it can be shown by induction that

$$\mathbf{\Phi}_k(\mathbf{z},0,\mathbf{n})=\left(\varphi_k(\mathbf{z},\mathbf{n}),\ \eta\frac{\partial\varphi_k}{\partial w}(\mathbf{z},\mathbf{n})\right).$$

Thus $\eta\,\partial\varphi_k/\partial w$ converges uniformly (to $\eta\,\partial\varphi/\partial w$). We can define

$$\mathbf{\Phi}(\mathbf{n})=\mathbf{\Phi}(\mathbf{n},\mathbf{w})=\lim_{k\to\infty}\mathbf{\Phi}_k(\mathbf{z},\mathbf{a},\mathbf{n})=\left(\varphi(\mathbf{n}),\frac{\partial\varphi}{\partial w}(\mathbf{n})\right).\tag{4.6}$$

Our argument thus far serves to show that $\partial\varphi/\partial w_k$ exists and is a continuous function of $\mathbf{n}$ for $k=1,\dots,K$. If $\mathbf{f}^1,\dots,\mathbf{f}^N$ have Lipschitz-continuous second derivatives with respect to $\mathbf{z}$ and $\mathbf{w}$, then the mappings in our augmented iterated function system given by (4.5) have Lipschitz-continuous first derivatives. We can apply the same argument to this augmented system to show that $\mathbf{\Phi}$ as given by equation (4.6) can be differentiated with respect to a second (perhaps identical) parameter w'. But then

$$\frac{\partial}{\partial w'}\mathbf{\Phi}=\left(\frac{\partial\varphi}{\partial w'},\frac{\partial^2\varphi}{\partial w\,\partial w'}\right);$$

thus φ can be continuously differentiated twice with respect to w and w'. ■

Note that the continuous twice-differentiability of φ with respect to $\mathbf{w}$ implies that χ is continuously twice-differentiable with respect to $\mathbf{w}$. We are actually more interested in differentiating χ than φ.

5. Taking Derivatives with Respect to Parameters

We now take up the problem of calculating the first and second derivatives of $V(\mathbf{w})$ with respect to $w_1, \ldots, w_K$.

Proof of Theorem 2.2. Let us consider V as given by equation (2.2):

$$V(\mathbf{w}) = \int_D \|\mathbf{v}(\mathbf{x}, \mathbf{w}) - \mathbf{u}(\mathbf{x})\|^q \, d\mu(\mathbf{x}).$$

We make the change of variable $\mathbf{x} = \boldsymbol{\psi}(\mathbf{n})$. We thus have $\mathbf{v}(\mathbf{x}, \mathbf{w}) = \boldsymbol{\chi}(\mathbf{n}, \mathbf{w})$:

$$V(\mathbf{w}) = \int_\Sigma \|\boldsymbol{\chi}(\mathbf{n}) - \mathbf{u}(\boldsymbol{\psi}(\mathbf{n}))\|^q \, d\sigma(\mathbf{n})$$

$$= \int_\Sigma \left(\sum_{j=1}^J |\chi_j(\mathbf{n}) - u_j(\boldsymbol{\psi}(\mathbf{n}))|^q \right) d\sigma(\mathbf{n}),$$

where χ_j and u_j are the jth components of $\boldsymbol{\chi}$ and $\mathbf{u}$, respectively. Note that we do not show the dependence of $\boldsymbol{\chi}$ on $\mathbf{w}$ explicitly.

The first and second derivatives of χ_j are continuous with respect to $\mathbf{n}$ and are therefore bounded on Σ. We can therefore differentiate under the integral sign:

$$\alpha_k = \frac{\partial V}{\partial w_k} = \int_\Sigma \sum_{j=1}^J \left(q\xi |\chi_j(\mathbf{n}) - u_j(\boldsymbol{\psi}(\mathbf{n}))|^{q-1} \frac{\partial \chi_j}{\partial w_k}(\mathbf{n}) \right) d\sigma(\mathbf{n}),$$

where $\xi = \operatorname{sgn}(\chi_j(\mathbf{n}) - u_j(\boldsymbol{\psi}(\mathbf{n})))$. Also,

$$\beta_{kl} = \frac{\partial^2 V}{\partial w_k \, \partial w_l} = \int_\Sigma \sum_{j=1}^J \Bigg[q(q-1) |\chi_j(\mathbf{n}) - u_j(\boldsymbol{\psi}(\mathbf{n}))|^{q-2} \frac{\partial \chi_j}{\partial w_k}(\mathbf{n}) \frac{\partial \chi_j}{\partial w_l}(\mathbf{n})$$

$$+ q\xi |\chi_j(\mathbf{n}) - u_j(\boldsymbol{\psi}(\mathbf{n}))|^{q-1} \frac{\partial^2 \chi_j}{\partial w_k \, \partial w_l}(\mathbf{n}) \Bigg] d\sigma(\mathbf{n}).$$

According to the ergodic theorem of Elton [7], these integrals can be calculated as trajectory averages. Choose $\mathbf{n}^0 \in \Sigma$ and for $t = 1, 2, \ldots$ define $\mathbf{n}^{t+1} = s^t(\mathbf{n}^t)$ with probability $p^n(\psi(\mathbf{n}^t))$. Set

$$\hat{a}^t_{j;k} = \frac{\partial \chi_j}{\partial w_k}(\mathbf{n}^t),$$

$$\hat{b}^t_{j;kl} = \frac{\partial^2 \chi_j}{\partial w_k \, \partial w_l}(\mathbf{n}^t).$$

Then with probability one,

$$\alpha_k = \lim_{T \to \infty} \frac{1}{T} \sum_{t=1}^T \sum_{j=1}^J q\xi |y^t_j - u_j(\mathbf{x}^t)|^{q-1} \hat{a}^t_{j;k}, \tag{5.1}$$

$$\beta_{kl} = \lim_{T \to \infty} \frac{1}{T} \sum_{t=1}^T [q(q-1)|y^t_j - u_j(\mathbf{x}^t)|^{q-2} \hat{a}^t_{j;k} \hat{a}^t_{j;l} + q\xi |y^t_j - u_j(\mathbf{x}^t)|^{q-1} \hat{b}^t_{j;kl}]. \tag{5.2}$$

Note the similarity of these equations to (2.4) and (2.5).

We now have the problem of finding $\hat{a}^t_{j;k}$ and $\hat{b}^t_{j;kl}$. These can be obtained from the functional equation (4.4):

$$\boldsymbol{\chi} \circ s^n = \mathbf{h}^n \circ \boldsymbol{\varphi}.$$

Let us set $\mathbf{x}^t = \boldsymbol{\psi}(\mathbf{n}^t)$, $\mathbf{y}^t = \boldsymbol{\chi}(\mathbf{n}^t)$, and $\mathbf{z}^t = \boldsymbol{\varphi}(\mathbf{n}^t) = (\mathbf{x}^t, \mathbf{y}^t)$. We then have, when $\mathbf{n}^{t+1} = s^n(\mathbf{n}^t)$,

$$y_j^{t+1} = h_j^n(\mathbf{x}^t, \mathbf{y}^t),$$

where h_j^n and y_j are the jth components of $\mathbf{h}^n$ and $\mathbf{y}$, respectively. Note $\hat{a}^t_{j;k} = \partial y_j^t/\partial w_k$ and $\hat{b}^t_{j;kl} = \partial y_j^t/\partial w_k\, \partial w_l$. Differentiating both sides of this equation with respect to $\mathbf{w}_k$ and applying the chain rule, we have

$$\hat{a}^{t+1}_{j;k} = \sum_{i=1}^{J} \frac{\partial h_j^n}{\partial y_i}(\mathbf{z}^t)\hat{a}^t_{i;k} + \frac{\partial h_j^n}{\partial w_k}(\mathbf{z}^t). \tag{5.3}$$

Another differentiation yields

$$\begin{aligned}\hat{b}^{t+1}_{j;kl} = {} & \sum_{i=1}^{J}\sum_{m=1}^{J} \frac{\partial^2 h_j^n}{\partial y_i\, \partial y_m}(\mathbf{z}^t)\hat{a}^t_{i;k}\hat{a}^t_{m;l} + \sum_{i=1}^{J} \frac{\partial^2 h_j^n}{\partial y_i\, \partial w_l}(\mathbf{x}^t, \mathbf{y}^t)\hat{a}^t_{i;k} \\ & + \sum_{i=1}^{J} \frac{\partial^2 h_j^n}{\partial y_i\, \partial w_k}(\mathbf{z}^t)\hat{a}^t_{i;l} + \frac{\partial^2 h_j^n}{\partial w_k\, \partial w_l}(\mathbf{z}^t) + \sum_{i=1}^{J} \frac{\partial h_j^n}{\partial y_i}(\mathbf{z}^t)\hat{b}^t_{i;kl}.\end{aligned} \tag{5.4}$$

These are essentially equations (2.2) and (2.3). We now show that they can be used to calculate $\hat{a}^t_{j;k}$ and $\hat{b}^t_{j;kl}$. Choose $a^0_{j;k}$ and $b^0_{j;kl}$ arbitrarily as approximations to $\hat{a}^0_{j;k}$ and $\hat{b}^0_{j;kl}$. For $t = 1, 2, \ldots$, calculate $a^t_{j;k}$ and $b^t_{j;kl}$ iteratively from equations (2.2) and (2.3). Let $\varepsilon^t_{j;k} = \hat{a}^t_{j;k} - a^t_{j;k}$ and $\zeta^t_{j;kl} = \hat{b}^t_{j;kl} - b^t_{j;kl}$ be the errors in these approximations. Then, subtracting (2.2) from (5.3), we have

$$\varepsilon^{t+1}_{j;k} = \sum_{i=1}^{J} \frac{\partial h_j^n}{\partial y_i}(\mathbf{z}^t)\varepsilon^t_{i;k}.$$

The fact that $\mathbf{f}^n$ is contracting with constant λ implies that, for fixed $\mathbf{x}$, $\mathbf{h}^n(\mathbf{x}, \mathbf{y})$ is contracting in $\mathbf{y}$ with constant λ. It follows that $\|\mathbf{Dh}^n\| \le \lambda$, where $\mathbf{Dh}^n$ is the derivative matrix of $\mathbf{h}^n$ with respect to $\mathbf{y}$. Thus $\varepsilon^t_{j;k}$ goes to zero at the rate of λ^t. Moreover, subtracting (2.3) from (5.4), we have

$$\begin{aligned}\zeta^{t+1}_{j;kl} = {} & \sum_{i=1}^{J}\sum_{m=1}^{J} \frac{\partial^2 h_j^n}{\partial y_i\, \partial y_m}(\mathbf{z}^t)(\hat{a}^t_{i;k}\hat{a}^t_{m;l} - a^t_{i;k}a^t_{m;l}) + \sum_{i=1}^{J} \frac{\partial^2 h_j^n}{\partial y_i\, \partial w_l}(\mathbf{z}^t)\varepsilon^t_{i;k} \\ & + \sum_{i=1}^{J} \frac{\partial^2 h_j^n}{\partial y_i\, \partial w_k}(\mathbf{z}^t)\varepsilon^t_{i;l} + \sum_{i=1}^{J} \frac{\partial h_j^n}{\partial y_i}(\mathbf{z}^t)\zeta^t_{i;kl}.\end{aligned}$$

Note that $(\hat{a}^t_{i;k}\hat{a}^t_{m;l} - a^t_{i;k}a^t_{m;l}) = \varepsilon^t_{i;k}\hat{a}^t_{m;l} + \varepsilon^t_{m;l}a^t_{i;k}$. Thus $\zeta^t_{j;kl}$ also goes to zero at the rate of λ^t. We can therefore substitute $a^t_{j;k}$ and $b^t_{j;kl}$ for $\hat{a}^t_{j;k}$ and $\hat{b}^t_{j;kl}$ in equations (5.1) and (5.2) without affecting the value of the average. ■

6. Affine Maps in the Plane

The treatment of a special case such as system consisting of affine maps proves to be considerably simpler than the general case outlined so far, because many of the derivatives and second derivatives are zero. We now consider the problem of approximating a function $u:[0,1]\to\mathbf{R}$ by a function v generated by an iterated function system consisting of N affine shear mappings in the plane; thus $L=J=1$.

We first choose constant probabilities $p^1,\ldots,p^N$. Our maps $g^1,\ldots,g^N:\mathbf{R}\to\mathbf{R}$ are then given by

$$g^n(x)=p^nx+\sum_{m<n}p^m.$$

We have specified that the probabilities, and hence the mappings g^n, shall be fixed. The functions $h^1,\ldots,h^N:\mathbf{R}^2\to\mathbf{R}$ are given by

$$h^n(x,y)=w_ny+w_{n+N}+w_{n+2N}x.$$

Thus there are altogether $K=3N$ parameters.

We therefore have

$$\frac{\partial h^n}{\partial y}=w_n,$$

$$\frac{\partial h^n}{\partial w_k}=\delta_{kn}y+\delta_{k,n+N}+\delta_{k,n+2N}x,\qquad k=1,\ldots,3N.$$

Equation (2.2) therefore becomes

$$a_k^{t+1}=w_na_k^t+\delta_{kn}y^t+\delta_{k,N+n}+\delta_{k,2N+n}x^t,\qquad k=1,\ldots,3N;$$

when $y^{t+1}=h^n(x^t,y^t)$.

We also have

$$\frac{\partial^2h^n}{\partial y^2}=\frac{\partial^2h^n}{\partial w_k\,\partial w_l}=0;$$

$$\frac{\partial^2h^n}{\partial y\,\partial w_k}=\delta_{kn}.$$

Equation (2.3) therefore becomes

$$b_{kl}^{t+1}=\delta_{kn}a_n^t+\delta_{ln}a_n^t+\omega_nb_{kl}^t,\qquad k,l=1,\ldots,3N.$$

In the case $q=2$, equations (2.4) and (2.5) become

(6.1) $$\alpha_k=\lim_{T\to\infty}\frac{1}{T}\sum_{t=1}^{T}2(y^t-u(x^t))a_k^t,$$

(6.2) $$\beta_{kl}=\lim_{T\to\infty}\frac{1}{T}\sum_{t=1}^{T}2[a_k^ta_l^t+(y^t-u(x^t))b_{kl}^t].$$

Finally, to determine the next set of parameter values $(w_1',\ldots,w_{3N}')$, we solve the system of $3N$ linear equations:

(6.3) $$0=\alpha_k+\sum_{l=1}^{3N}\beta_{kl}\,\Delta w_l,$$

for $k=1,\ldots,3N$ and let $w_k'=w_k+\Delta w_k$.

We have implemented these formulas in a computer program, using $n=3$; thus there are nine parameters. The limits in equations (6.1) and (6.2) are approximated by taking the value 500 or 1000 for T. Figures 1–5 show successive approximations to the target function $u(x)=\sin 2\pi x$. After three iterations, the calculated value of V has decreased from 1.08582 to 0.01215 and $\|\nabla V\|$ has decreased from 1.1921 to 0.0139. Under further iteration the calculated value for $\|\nabla V\|$ continues to creep down, but it seems apparent that essentially the best approximation has been reached. At this level the error due to the finiteness of T probably becomes significant.

Figures 6–11 show the process as applied to the target function $u(x)=9x(1-x)$. After five iterations, the calculated value of V has decreased from 0.1873 to 0.0000071 and $\|\nabla V\|$ has decreased from 1.6256 to 0.00224. Note the symmetry breaking due to the random element.

The first example demonstrates that we need not begin with a value $\mathbf{w}$ which is very close to $\mathbf{w}^*$ in order for the iterates to converge to $\mathbf{w}^*$. However, other examples not included here show that for many initial values of $\mathbf{w}$, the iterates eventually leave the region of parameter space where the system has an attracting invariant measure, and our method is then no longer applicable. Given the complicated dynamics and structure of basin boundaries for Newton's method (as discussed, for example, in Hurley and Martin [9], Saari and Urenko [11],

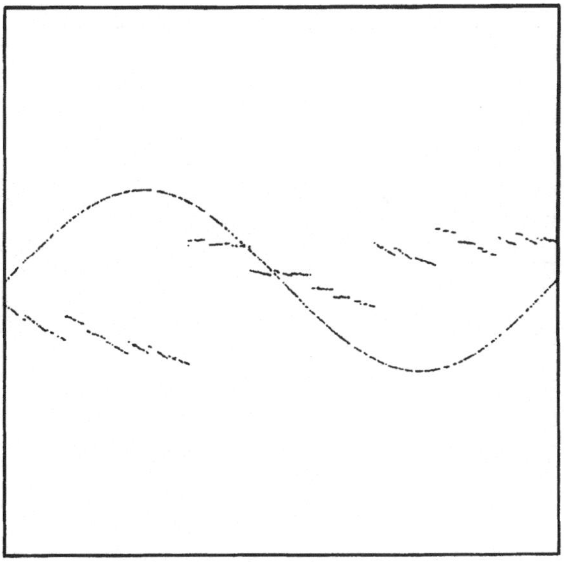

Iteration 0: $V=1.086$, $\|\nabla V\|=1.1921$

$w_1 = 0.2$ $w_2=-0.2$ $w_3= 0.3$
$w_4=-0.2$ $w_5= 0.4$ $w_6= 0.5$
$w_7=-0.8$ $w_8=-0.6$ $w_9=-0.2$

Fig. 1

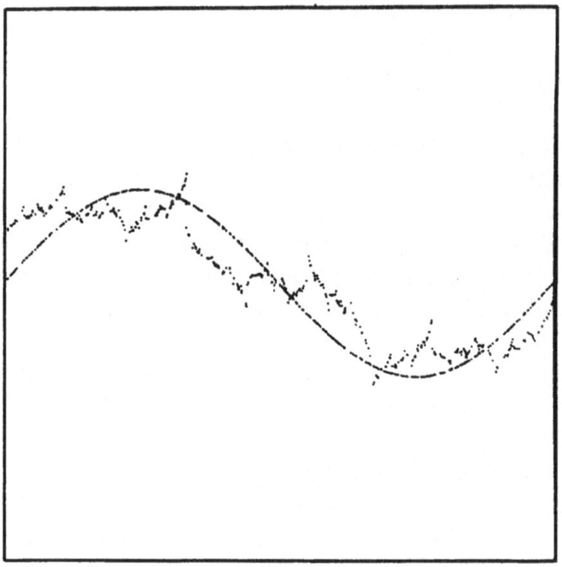

Iteration 1: $V = 0.0928$, $\|\nabla V\| = 0.2392$		
$w_1 = 0.4428$	$w_2 = -0.5079$	$w_3 = 0.4696$
$w_4 = 0.2580$	$w_5 = 0.8440$	$w_6 = -1.395$
$w_7 = 0.9698$	$w_8 = -1.725$	$w_9 = 1.345$

Fig. 2

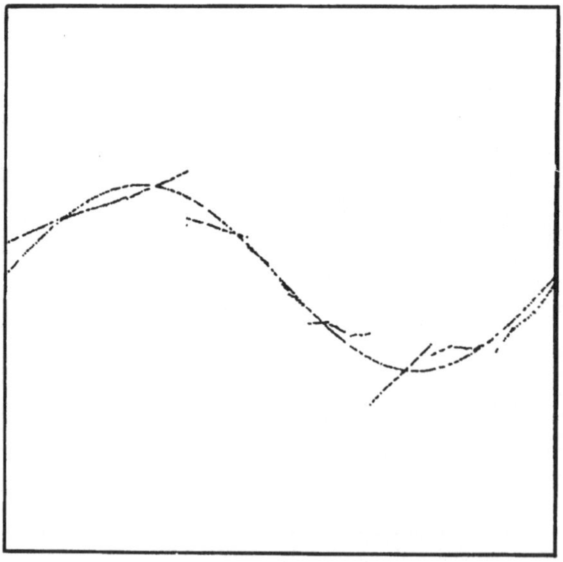

Iteration 2: $V = 0.0190$, $\|\nabla V\| = 0.0406$		
$w_1 = 0.0305$	$w_2 = -0.2188$	$w_3 = 0.2320$
$w_4 = 0.3360$	$w_5 = 0.5524$	$w_6 = -1.444$
$w_7 = 0.8182$	$w_8 = -1.132$	$w_9 = 1.393$

Fig. 3

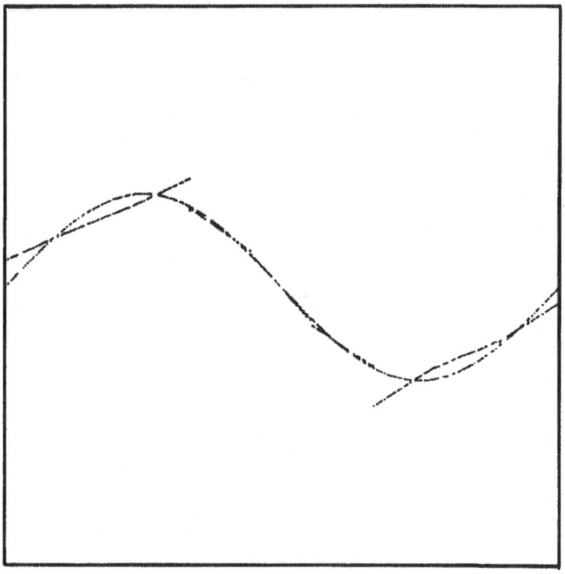

Iteration 3: $V = 0.0121$, $\|\nabla V\| = 0.0139$

$w_1 = 0.0101$	$w_2 = -0.1009$	$w_3 = 0.0477$
$w_4 = 0.2867$	$w_5 = 0.8200$	$w_6 = -1.286$
$w_7 = 0.8677$	$w_8 = -1.653$	$w_9 = 1.130$

Fig. 4

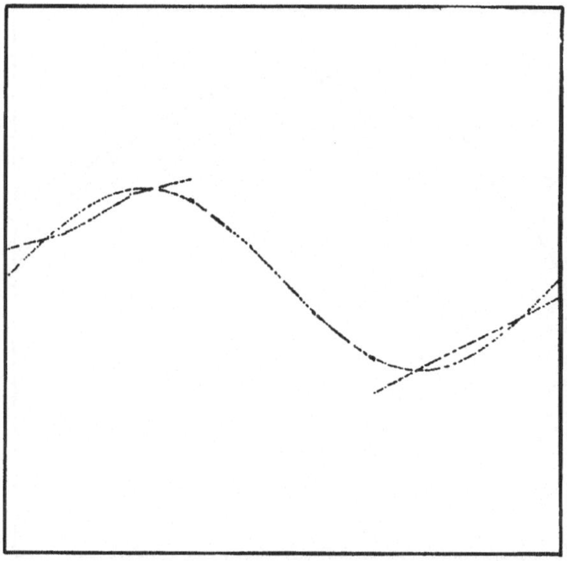

Iteration 10: $V = 0.0115$, $\|\nabla V\| = 0.000044$

$w_1 = -0.0732$	$w_2 = 0.0565$	$w_3 = 0.0102$
$w_4 = 0.3424$	$w_5 = 0.8802$	$w_6 = -1.251$
$w_7 = 0.7456$	$w_8 = -1.768$	$w_9 = 1.056$

Fig. 5

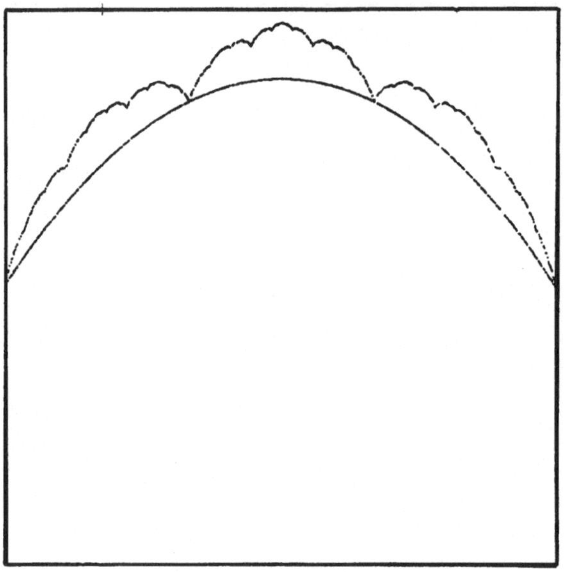

Iteration 0: $V = 0.1874$, $\|\nabla V\| = 1.626$

$w_1 = 0.3$	$w_2 = 0.3$	$w_3 = 0.3$
$w_4 = 0$	$w_5 = 2$	$w_6 = 2$
$w_7 = 2$	$w_8 = 0$	$w_9 = -2$

Fig. 6

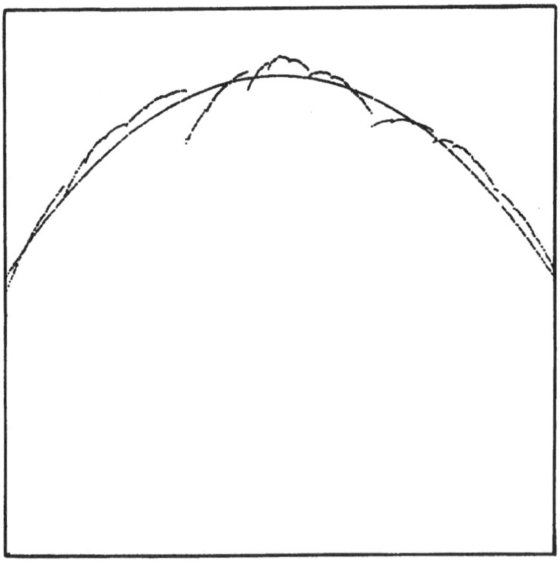

Iteration 1: $V = 0.0214$, $\|\nabla V\| = 0.2532$

$w_1 = 0.2052$	$w_2 = 0.3264$	$w_3 = 0.2181$
$w_4 = -0.1411$	$w_5 = 1.526$	$w_6 = 1.698$
$w_7 = 2.209$	$w_8 = 0.2706$	$w_9 = -1.635$

Fig. 7

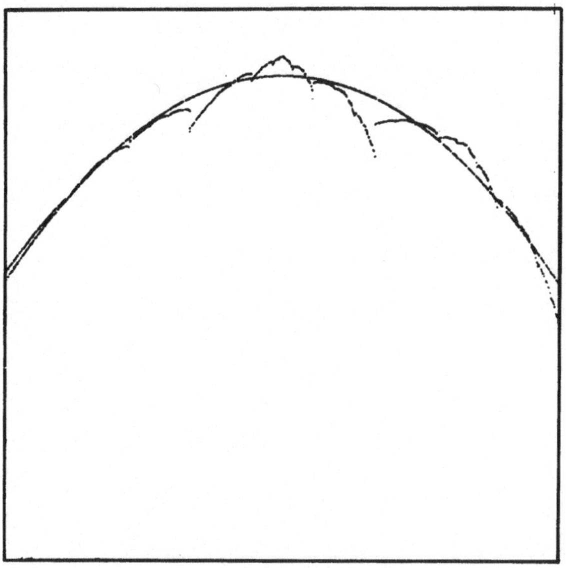

Iteration 2: $V = 0.0165$, $\|\nabla V\| = 0.0747$		
$w_1 = 0.1169$	$w_2 = 0.3702$	$w_3 = 0.3164$
$w_4 = 0.0814$	$w_5 = 1.627$	$w_6 = 1.686$
$w_7 = 1.831$	$w_8 = -0.1312$	$w_9 = -1.983$

Fig. 8

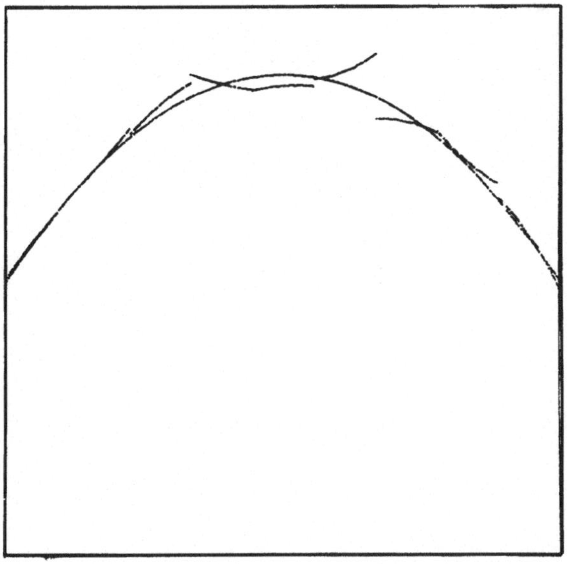

Iteration 3: $V = 0.0117$, $\|\nabla V\| = 0.0623$		
$w_1 = 0.0887$	$w_2 = -0.1136$	$w_3 = 0.2124$
$w_4 = -0.0332$	$w_5 = 2.258$	$w_6 = 1.765$
$w_7 = 2.198$	$w_8 = 0.2192$	$w_9 = -1.845$

Fig. 9

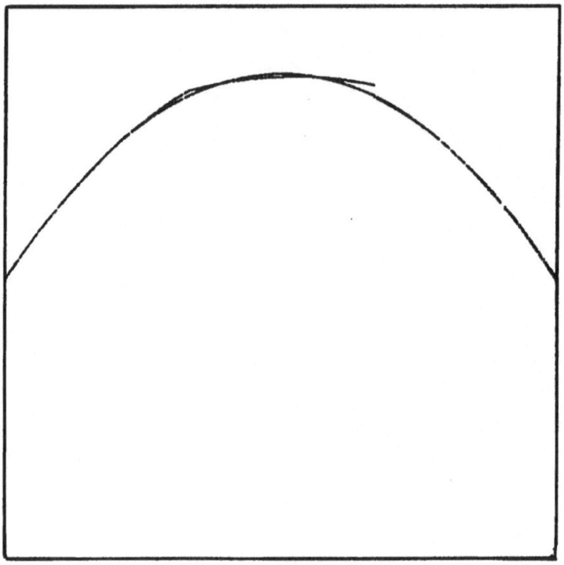

Iteration 4: $V = 0.00072$, $\|\nabla V\| = 0.0194$

$w_1 = 0.0924$	$w_2 = 0.0573$	$w_3 = 0.1091$
$w_4 = 0.0155$	$w_5 = 2.057$	$w_6 = 1.986$
$w_7 = 2.041$	$w_8 = 0.0633$	$w_9 = -1.952$

Fig. 10

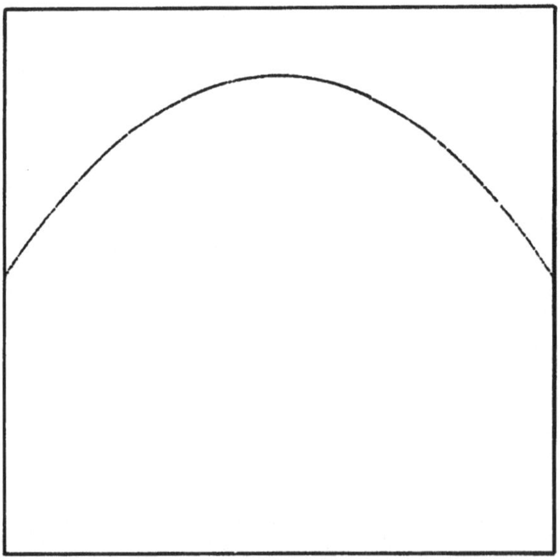

Iteration 5: $V = 0.0000071$, $\|\nabla V\| = 0.00224$

$w_1 = 0.1109$	$w_2 = 0.1057$	$w_3 = 0.1151$
$w_4 = -0.0617$	$w_5 = 2.007$	$w_6 = 1.993$
$w_7 = 2.003$	$w_8 = 0.00362$	$w_9 = -1.996$

Fig. 11

and Wong [15]), it seems unlikely that there is any simple method to determine in advance whether the iterates of a given initial $\mathbf{w}$ converge to $\mathbf{w}^*$ or not. One naive, but perhaps practical, way to cope with this is simply to try initial conditions at random until one converges. An alternative approach is to use a parameter value obtained from some sort of interpolation for the first value of $\mathbf{w}$, as we did in our parabola example.

Another potential difficulty is that the matrix of second derivatives $[\beta_{kl}]$ of V might be singular. We have observed this in practice only when some of the parameters are set to zero (we therefore refrain from choosing zero values for some parameters in our initial condition) or in treating the case $q=1$. Even if $[\beta_{kl}]$ is singular, it might still be possible to apply our method if the system of equations (6.3) has a solution.

We note also that the approximating function thus obtained may not be continuous. If a continuous approximating function is desired, it can be obtained by appropriately expressing $(w_1, \ldots, w_{3N})$ in terms of $(W_1, \ldots, W_{2N+1})$:

$$w_k = W_k, \qquad k = 1, \ldots, 2N,$$

$$w_{3N} = W_{2N+1},$$

$$w_{k+2N} = W_{k+1}\frac{W_{N+1}}{1-W_1} + W_{k+N+1} - W_k\frac{W_{2N}+W_{2N+1}}{1-W_N} - W_{k+N},$$

$k = 1, \ldots, N-1$. The algorithm can then be implemented using $\partial h^n/\partial W_k$, etc.

Similarly, the capacity of the graph of the approximating function can be adjusted by restricting the parameters; for example, to conform with the formulas in [6] or [8].

Acknowledgment. This research was supported by the Naval Academy Research Council.

References

1. M. F. Barnsley (1986): *Making chaotic dynamical systems to order.* In: Chaotic Dynamics and Fractals (M. F. Barnsley, S. G. Demko, eds.). New York: Academic Press, 53–68.
2. M. F. Barnsley, S. G. Demko (1985): *Iterated function systems and the global construction of fractals.* Proc. Roy. Soc. London Ser. A, **399**:243–275.
3. M. F. Barnsley, S. G. Demko, J. H. Elton, J. S. Geronimo (preprint): *Invariant measures for Markov processes arising from iterated function systems with place-dependent probabilities.*
4. M. F. Barnsley, S. G. Demko, J. H. Elton, J. S. Geronimo (preprint): *On the attractiveness of invariant measures for iterated function systems.*
5. M. F. Barnsley, J. H. Elton (preprint): *Stationary attractive measures for a class of Markov chains arising from function iteration.*
6. M. F. Barnsley, J. H. Elton, D. P. Hardin, P. R. Massopoust (preprint): *Hidden-variable fractal interpolation functions.*
7. J. H. Elton (1987): *An ergodic theorem for iterated maps.* Ergodic Theory Dynamical Systems, **7**:481–488.
8. D. P. Hardin, P. R. Massopoust (1986): *The capacity for a class of fractal functions.* Comm. Math. Phys., **105**:455–460.
9. M. Hurley, C. F. Martin (1984): *Newton's algorithm and chaotic dynamical systems.* SIAM J. Math. Anal., **15**:238–252.

10. B. B. MANDELBROT (1982): The Fractal Geometry of Nature. San Francisco: Freeman.
11. D. G. SAARI, J. B. URENKO (1984): *Newton's method, circle maps, and chaotic motion.* Amer. Math. Monthly, **91**:3–17.
12. W. D. WITHERS (1986): *Calculation of Taylor series for Julia sets in powers of a parameter.* In: Chaotic Dynamics and Fractals (M. F. Barnsley, S. G. Demko, eds.). New York: Academic Press, pp. 203–213.
13. W. D. WITHERS (submitted): *Differentiability with respect to parameters of average values in probabilistic contracting dynamical systems.* Ergodic Theory Dynamical Systems.
14. W. D. WITHERS (1987): *Calculating derivatives with respect to parameters in iterated function systems.* Physica, **28D**:206–214.
15. S. WONG (1984): *Newton's method and symbolic dynamics.* Proc. Amer. Math. Soc., **91**:245–253.

W. D. Withers
Department of Mathematics
United States Naval Academy
Annapolis
Maryland 21402
U.S.A.

CONSTRUCTIVE APPROXIMATION

Instructions to Authors

General Information

Manuscripts for this journal must be in English and should be typed on only one side of the sheet of paper, double spaced, with two-inch margins all around. All elements of formulae, also, should be typewritten whenever possible.

Each figure (line drawings and graphs) should be submitted on a separate sheet of paper; each diagram should be drawn precisely in India ink, using lines of uniform width. The figures should be drawn at twice the size desired: the maximum area for the reduced figures is $4\frac{3}{4} \times 7\frac{7}{8}$ in., or 122 by 194 mm. All diagrams should be numbered consecutively, and their first mention in the text should be noted in the left-hand margin. Legends should be listed consecutively on a single sheet of paper.

An abstract of the article should be included on the title page along with AMS subject classifications and key words and phrases.

Footnotes other than those referring to the title should be avoided. If they are essential, they should be numbered consecutively and placed at the foot of the page to which they refer.

The references to the literature should be listed alphabetically at the end of the manuscript. For journals, the following information should appear: names (including initials or first names) of all authors, year of publication, full title of paper, journal name, volume, and pages. Books cited should list author(s), year, full title, edition, publisher, and place of publication. Examples are:

1. D. Jackson (1912): *On the approximation by trigonometric sums and polynomials.* Trans. Amer. Math. Soc., **13**:491–515.

2. G. G. Lorentz (1966): Approximation of Functions. New York: Holt, Rinehart and Winston.

Authors are encouraged to use the symbol ":=" to denote definitions that appear in the manuscript.

Marking the Manuscript

In text, words or phrases requiring italic type should be underlined. Only the end of a proof should be indicated by □.

All letters in formulae, as well as single letters in the text, are automatically printed italic and therefore require no underlining. Please use the following color code for marking:

Greek letters: red
Gothic letters: blue
Script letters: green
Regular (Roman) letters: yellow
Special Roman letters: wavy blue
Boldface: brown
Italics: underline in pencil

Please use pencil for all editorial marks, as the compositor is likely to interpret an ink mark as an element to be printed. If its appearance is ambiguous, underline a capital letter three times. In complicated sections of the text, mark superscripts and subscripts with a caret: $\mathrm{b}^{\vee 1}$, $\mathrm{b}_{\widehat{\mathrm{j}}}$. Ordinarily subscripts appear below superscripts; they should be typed this way, and marked if variation is desired.

The following elements are often confused and should be identified by the appropriate previously discussed measures or by a circled word or words explaining the element:

$$\cup, \mathrm{u}, U; {}^\circ, o, O, 0; \times, x, X, \kappa; \vee, v, \upsilon, \nu;$$
$$\theta, \Theta; \phi, \varphi, \Phi, \emptyset, \varnothing; \psi, \Psi; \varepsilon, \epsilon, \in; a', a^1;$$
$$c, C; e, l; I, J; k, K; o, O; p, P; s, S;$$
$$u, U; v, V; w, W; x, X; z, Z.$$

The letter O and the number 0, and the letter l and the number 1 occur frequently. Each appearance of one or zero should be identified. The number 1 should be written with a hook and baseline bar. Distinguish between the indefinite article a and the italic letter *a* used in a mathematical context.

Miscellaneous

Manuscripts submitted according to these instructions will reduce considerably the cost of charges to the author in proofs. The only corrections to be made in proofs are typographical errors. Should the author wish to correct stylistic or factual errors in proofs, he must absorb the cost.

Papers for publication should be submitted in triplicate to *Constructive Approximation*, Department of Mathematics, University of South Florida, Tampa, FL 33620. The author's address must be clearly indicated, and the editorial office must be notified immediately of a change of address.

CONSTRUCTIVE APPROXIMATION

Editors-in-Chief

Ronald A. DeVore
Department of Math and Statistics
University of South Carolina
Columbia, South Carolina 29208

Edward B. Saff
Department of Mathematics
University of South Florida
Tampa, Florida 33620

Editorial Board

J. Milne Anderson
Mathematics Department
University College
London WC1E 6BT, England

Hubert Berens
Mathematisches Institut
Universität Erlangen-Nürnberg
Bismarckstrasse $1\frac{1}{2}$
D-8520 Erlangen, West Germany

Carl de Boor
Mathematics Research Center
610 Walnut Street
Madison, Wisconsin 53705

Dietrich Braess
Institut für Mathematik
Ruhr-Universität Bochum
4630 Bochum 1, West Germany

Wolfgang Dahmen
Institut für Mathematik III
Freie Universität Berlin
Arnimallee 2–6
1000 Berlin (West) 33, Germany

Dieter Gaier
Mathematisches Institut
Universität Giessen
Arndstrasse 2
D 63 Giessen, West Germany

William B. Gragg
Department of Mathematics
University of Kentucky
Lexington, Kentucky 40506

Klaus Höllig
Universität Stuttgart
Mathematisches Institut A
D-7000, Stuttgart-80
Pfaffenwaldring-57
West Germany

Mourad Ismail
Department of Mathematics
University of South Florida
Tampa, Florida 33620

Tom H. Koornwinder
Centre for Mathematics and Computer Science
P.O. Box 4079
1009 AB Amsterdam
The Netherlands

George G. Lorentz
Department of Mathematics
RLM 8–100
University of Texas
Austin, Texas 78712

Charles A. Micchelli
IBM Research Center
Box 218
Yorktown Heights
New York 10598

Paul Nevai
Department of Mathematic
The Ohio State University
Columbus, Ohio 43210

Allan Pinkus
Department of Mathematic
Technion
32000 Haifa, Israel

M. J. D. Powell
Department of Applied Mathematics and Theoretical Physics
University of Cambridge
Silver Street
Cambridge CB3 9EW
England

Larry L. Schumaker
Department of Mathematic
Vanderbilt University
Nashville
Tennessee 37235

Richard S. Varga
Institute for Computational Mathematics
Kent State University
Kent, Ohio 44242

SCOPE OF JOURNAL

Constructive Approximation is an international mathematics journal dedicated to Appro tions and Expansions and related research in Computation; Function Theory; Func Analysis; Interpolation Spaces and Interpolation of Operators; Numerical Analysis; Spa Functions; Special Functions, and applications.

SUBMISSION OF MANUSCRIPTS

Papers intended for publication in *Constructive Approximation* should be submitt triplicate to E. B. Saff, Editor-in-Chief, Department of Mathematics, University of South Fl Tampa, FL 33620. All manuscripts must be written in English and should contain an abs key words and phrases; and current AMS MOS classification numbers.